The
ASVAB Tutor's
Arithmetic
Reasoning
Study Guide
by
Julie A. Hyers

The ASVAB Tutor's Arithmetic Reasoning Study Guide
Julie A. Hyers © 2020

The Arithmetic Reasoning section of the ASVAB consists of math word problems.

The book has broken the section into 15 subtopics and consists of:

- 15 introductory lessons covering each subtopic
- 15 pre-tests with explanations of answers
- 15 videos solving the pre-test questions (online access with purchase of videos)
- 15 post-tests with explanations of answers

15 Arithmetic Reasoning Subtopics:
1. Basic Operations
2. Decimals
3. Fractions
4. Mixed Numbers
5. Percentage
6. Percent, Fraction, Decimal Conversions
7. Proportions
8. Ratios
9. Tax, Tip, Commission and Overtime
10. Interest Rate
11. Patterns and Sequences
12. Probability and Consecutive Integers
13. Distance and Rate
14. Miles per Gallon and Using Algebra to Solve Word Problems
15. Measurement

Arithmetic Reasoning Post-Tests pages 120-182

*Not responsible for typographical errors.

Reviews and Pre-Tests

Basic Operations Review

Terms to Know for Basic Operations
- **Sum** - answer in addition
- **Difference** - answer in subtraction
- **Product** - answer in multiplication
- **Quotient** - answer in division
- **Remainder** - the number that is left over when a number does not divide evenly into another number
- **Divisor** - a number that divides into another number
- **Dividend** - a number that is to be divided by another number

Addition
Add.
Carry if necessary.

```
    9,367
+   2,453
   11,820
```

Subtraction
Subtract.
Borrow if necessary.

```
   20,000
-  15,982
    4,018
```

Multiplication
Make sure to be strong with your times tables.
Every second counts on this test!

Multiplication by a one-digit number

```
    243
  x   7
  1,701
```

Multiplication by a two-digit number
First multiply by the ones place.
Then multiply by the tens place.
Add the results.

```
    463
  x 28
  3,704
+ 9,260
 12,964
```

Here is the work broken down into steps.
The first step is to multiply by the ones place.

```
    463
  x   8
  3,704
```

The next step is to multiply by the tens place.
Write one zero in the last place, then multiply the other numbers.

```
    463
  x  20
  9,260
```

Then add up the two results.

```
    3,704
  + 9,260
   12,964
```

Multiplication by a three-digit number.

```
      567
  x   348
    4,536
   22,680
+ 170,100
  197,316
```

Here is the work broken down into steps.
The first step is to multiply by the ones place.

```
    567
  x   8
  4,536
```

The next step is to multiply by the tens place.
Write one zero in the last place, then multiply the other numbers.

```
    567
  x  40
 22,680
```

Then multiply by the hundreds place.
Write two zeros in the last two places, then multiply the other numbers.

```
        567
      x 300
    170,100
```

Finally, add up the three results.

```
      4,536
     22,680
   + 170,100
    197,316
```

Division

Make sure to be strong with your times tables.
Every second counts on this test!
A way to remember the steps of long division is:
<u>D</u>oes <u>M</u>om <u>S</u>erve <u>C</u>heese <u>B</u>urgers?
The steps of long division are:
<u>D</u>ivide
<u>M</u>ultiply
<u>S</u>ubtract
<u>C</u>heck
<u>B</u>ring down

Division by a one-digit number

Divide by the digit, one step at a time.

```
        1,189 remainder 1

    7 ⟌ 8,324
       -7
       13
       -7
       62
      -56
       64
      -63
        1
```

Division by a two-digit number

Divide by the two-digit number, step by step.

```
             293 remainder 6
        ┌─────────
   23  )  6,745
        -46
         214
        - 207
           75
          -69
            6
```

Division by a three-digit number

Divide the three digits in step by step.
Three-digit division can be a bit trickier.

To make it easier, use estimation. Estimate by figuring out how many times the first digit or the first two digits go into the number to make the division easier.

In this case, it is difficult to figure out how many times 354 goes into 789.
To estimate, guess how many times 35 would go into 78.
Or estimate how many times 3 would go into 7.
The answer would be 2.
Then figure out 354 x 2 to begin to solve.

```
             223 remainder 41
        ┌─────────
  354  )  78,983
         -708
          818
         -708
          1103
        - 1062
            41
```

Arithmetic Reasoning

Basic Operations - Pre-Test

1.   8,654
   +   4,978

2.   30,000
   - 18,754

3.   167
    x   8

4.   342
    x   67

5.   427
      x 268

6.   852 ÷ 8 =

7.   4,833 ÷ 75 =

8.   87,653 ÷ 248 =

Arithmetic Reasoning

Basic Operations - Pre-Test with Answers

1. 8,654
    + 4,978
    ─────
    13,632

2. 30,000
    -18,754
    ─────
    11,246

3. 167
    x 8
    ────
    1,336

4. 342
    x 67
    ────
    2394
    + 20520
    ─────
    22,914

5.   427
     x 268
     3416
   25620
 + 85400
 114,436

6. $852 \div 8 =$

       106 remainder 4

  8 ⟌ 852
     -8
     05
    - 0
     52
    -48
      4

7. $4,833 \div 75 =$

       64 remainder 33

  75 ⟌ 4833
    - 450
     333
    - 300
     33

8. $87,653 \div 248 =$

353 remainder 109

$$
\begin{array}{r}
248 \overline{)\ 87,653} \\
-\ 744 \\
\hline
1325 \\
-\ 1240 \\
\hline
853 \\
-\ 744 \\
\hline
109
\end{array}
$$

Decimal Review

Adding Decimals
Make sure you line up the decimals. Then add.
If a number does not have a decimal point, the decimal point is at the end.
Always line up the decimal points. Add zeros after the decimal point if necessary.

$3.45 + 0.0056 + 210 =$

```
 210.0000
   3.4500
+  0.0056
 213.4556
```

Subtracting Decimals
Make sure you line up the decimals. Then subtract.
If a number does not have a decimal point, the decimal point is at the end.
Always line up the decimal points. Add zeros after the decimal point if necessary.

$1,567.2 - 0.0451 =$

```
  1,567.2000
-      0.0451
  1,567.1549
```

Multiplying Decimals
There is no need to line up decimals.
Multiply normally.
At the end, count up how many digits appear after the decimal point in the question.
Add up the total number of digits after the decimal point.
In the solution, start on the right side of the answer and move your decimal point to the left that many spaces.

```
    .124          3 places after the decimal point
  x .35        +  2 places after the decimal point
    620           5 places total
+ 3720
0.04340
```

In the answer, move the decimal point 5 places to the left.
When you move the decimal place 5 places to the left on 4340, the answer is 0.04340

14

Dividing Decimals

If the number you are dividing by (the divisor) has a decimal, you must move it to the right, after the number. However many places you move the decimal to the right in the divisor, you must move the same number of places to the right in the dividend (the number that it is being divided into).

$320 \div 0.06 =$

$$0.06 \overline{)\ 320}$$

In dividing by 0.06, you must move the decimal 2 places to the right to become a whole number, 6.

Then move the decimal 2 places to the right in 320 to become 32000.

Then divide.

$32000 \div 6 =$

```
         5333.3
    6 ) 32000.0
       -30
         20
        -18
          20
         -18
           20
          -18
            20
           -18
             2
```

If you are dividing decimals and the dividend (the number that you are dividing into) has a decimal, bring up the decimal and divide normally.

There is no need to move the decimal.

$9811.5 \div 93 = 105.5$

```
           105.5
    93 ) 9811.5
        -93
          51
         - 0
          511
         -465
           465
          -465
             0
```

Julie A. Hyers ©2020

If you are dividing decimals and the dividend (the number that you are dividing into) and the divisor (the number that you are dividing by) both have decimals, move the decimal over after the last number of the divisor. Then move it the same number of spaces in the dividend.

Example:
1,189.41 ÷ 12.3

$$12.3 \overline{)\ 1,189.41}$$

12.3 has one number after the decimal point.
The decimal point needs to be moved one place to the right to become 123.
Since we moved the decimal point one place to the right in 12.3, we need to move the decimal one place to the right in 1,189.41 to become 11,894.1
The problem becomes:

11894.1 ÷ 123 = 96.7

$$
\begin{array}{r}
96.7 \\
123 \overline{)\ 11,894.1} \\
-\ 1107 \\
\hline
824 \\
-\ 738 \\
\hline
861 \\
-\ 861 \\
\hline
0
\end{array}
$$

16

Arithmetic Reasoning

Decimals - Pre-Test

1.   Add decimals.
        $3.27 + 0.0058 + 400 =$

2.   Subtract decimals.
        $652.8 - 0.008 =$

3.   Multiply decimals.
        $400.25 \times .34 =$

4.   Divide decimals.
        $180 \div .45 =$

Arithmetic Reasoning

Decimals - Pre-Test with Answers

1. Add decimals.
    $3.27 + 0.0058 + 400 =$ Line up the decimal places before adding.

    ```
      400.0000
        3.2700
    +   0.0058
      403.2758
    ```

2. Subtract decimals.
    $652.8 - 0.008 =$ Line up the decimal places before subtracting.

    ```
    652.800
    - 0.008
    652.792
    ```

3. Multiply decimals.
    $400.25 \times .34 =$

    In this case, there are 2 digits after the decimal place in 400.25 and 2 digits after the decimal place in .34. $2 + 2 = 4$, and the decimal place in the answer needs to be moved 4 places to the left in the answer. 1360850 becomes 136.0850

    ```
        400.25
    x      .34
        160100
    + 1200750
      136.0850
    ```

4. Divide decimals.
    $180 \div .45 =$

    When dividing by a decimal, move the decimal over to the right so it is behind the number and then move the decimal the same number of places in the other number.

    ```
                      400
    0.45 ) 180  =  45 ) 18000
                      -180
                        00
                      -  0
                        00
                      -  0
                         0
    ```

Fraction Review

Numerator- top number in a fraction

Denominator- bottom number in a fraction

Reciprocal- a fraction is flipped over

Example: The reciprocal of 2/4 is 4/2.

The reciprocal of 3/5 is 5/3.

The reciprocal of 4 is 1/4 because 4 is 4/1, which becomes 1/4.

Changing a mixed number into an improper fraction

1   3/4 = 7/4

Multiply the whole number by the denominator and add the numerator to form the numerator and keep the same denominator.

Changing an improper fraction into a mixed number

11/9 = 1 2/9

Divide the denominator into the numerator. The answer becomes the whole number. The remainder becomes the numerator. Keep the same denominator.

Reducing Fractions

To find a final answer in a fraction problem, the answer must be in the most reduced form.

For example: 4/10 can be reduced since 4 and 10 can both be divided by 2.
4/10 would reduce to 2/5.
Remember if the numerator and denominator are even, you can divide the numerator and denominator by the number 2 to reduce the fraction.

For example: 3/6 can be reduced since 3 and 6 can both be divided by 3.
3/6 would reduce to 1/2.

Another example: 14/35 can be reduced since 14 and 35 can both be divided by 7.
14/35 would reduce to 2/5.

Multiple – a number that results when a number is multiplied by 1, 2, 3, and so on.

Example: The multiples of 6 are: 6, 12, 18, 24, 30, 36, etc.

Factor- a number that divides into another number without a remainder
Example: The factors of 12 are: 1, 2, 3, 4, 6, 12.

Least Common Multiple (LCM)
LCM is the smallest number that two numbers can be multiplied to in order to find a common denominator.

LCM of 6 and 9 is 18.
LCM of 5 and 12 is 60.
LCM is used to find a common denominator.

Greatest Common Factor (GCF)
GCF is the largest number that 2 numbers can be divided by.
Example: The GCF of 18 and 24 is 6.
GCF is used to reduce fractions.

Always make sure to reduce your final answer in fractions if able to be reduced.

Adding Fractions

In adding fractions, if you have a common denominator, just add the numerators.

Example: 2/4 + 1/4 = 3/4

4 is the common denominator. There is no need to find a common denominator.
Just add the numerators to solve.

In adding fractions, if you do not have a common denominator, you must find a common denominator. Then add the numerators, not the denominators.

The common denominator can be found by listing the multiples of the 2 denominators.

Example: 3/4 + 1/6 has no common denominator.

To find a least common denominator, list the multiples of the denominators 4 and 6.

4	6
8	12
12	18

It can be seen that the least common denominator would be 12.

Whatever you do to the denominator, you do to the numerator.

3/4 becomes 9/12 because 4 x 3 = 12.
Since we multiplied the denominator by 3, we need to multiply the numerator by 3.
3 x 3 = 9 leading to 9/12

1/6 becomes 2/12 because 6 x 2 =12.
Since we multiplied the denominator by 2, we need to multiply the numerator by 2.
1 x 2 = 2 leading to 2/12

It becomes 9/12 + 2/12 =
Then add the numerators, not the denominators.
9/12 + 2/12 = 11/12

If you are having a tough time finding a common denominator, you could always multiply the two denominators by each other to find a common denominator. This denominator might not be the least common denominator but would still work. If you multiplied 6 x 4, you would end up with a common denominator of 24.

Another way to add fractions is to multiply the denominators to find the denominator and cross multiply the numerators and denominators and then add them.

3/4 + 1/6

Multiply the denominators. 6 x 4 = 24
24 will be the denominator.

Cross-multiply 3 x 6 = 18
Cross-multiply 4 x 1 = 4
Add the 18 and 4 to get the numerator.
18 + 4 = 22

The answer is 22/24, which can be reduced to 11/12.

Subtracting Fractions
In subtracting fractions, if you have a common denominator, just subtract the numerators.

Example: 3/5 - 2/5 = 1/5
5 is the common denominator. There is no need to find a common denominator.
Just subtract the numerators to solve.
Do not subtract the denominators.

In subtracting fractions, if there is no common denominator, you must find a common denominator.
Example: 4/7 – 1/2

4/7 – 1/2 has no common denominator.

To find a common denominator, list the multiples of the denominators 7 and 2.

2	7
4	<u>14</u>
6	21
8	28
10	35
12	42
<u>14</u>	49

The least common denominator for 7 and 2 is 14.

Whatever you do to the denominator, you must do to the numerator.

4/7 becomes 8/14 because 7 x 2 = 14.
Since we multiplied the denominator by 2, we need to multiply the numerator by 2.
4 x 2 = 8 leading to 8/14
1/2 = 7/14 because 2 x 7 =14.
Since we multiplied the denominator by 7, we need to multiply the numerator by 7.
Then we need to multiply 1 x 7 = 7.

It becomes 8/14 – 7/14 =

Then subtract the numerators, not the denominators.
8/14 - 7/14 = 1/14

If you are having a tough time finding a common denominator, you could always multiply the two denominators by each other to find a common denominator.
This denominator might not be the least common denominator but would still work.
If you multiplied 7 x 2, you would end up with a common denominator of 14.
In this case, this is the least common denominator.
Another way to subtract fractions is to multiply the denominators to find the denominator and cross multiply the numerators and denominators and then subtract them.

4/7 - 1/2
Multiply the denominators.
7 x 2 = 14
14 will be the denominator.
Cross-multiply 4 x 2 = 8
Cross-multiply 7 x 1 = 7
Subtract the 8 and 7 to get the numerator.
8 - 7 = 1
The answer is 1/14.

Multiplying Fractions

Multiply the numerators and multiply the denominators straight across.
Multiply the top. Multiply the bottom.
Do NOT find a common denominator.

5/7 x 2/4 = 10/28
Reduce if necessary.
10/28 can be divided by 2 on the top and bottom, which equals 5/14.

Dividing Fractions

Keep the first fraction. Change the sign from division to multiplication and flip the second fraction (find its reciprocal).
Do NOT find a common denominator.

Keep, change, flip.

3/5 ÷ 2/3 =

3/5 x 3/2 = 9/10

Comparing Fractions

There are a few ways to compare fractions.

1.   Find a common denominator. Then compare.

        3/5 compared to 5/7

        List the multiples of 5 and 7.
        5   7
        10   14
        15   21
        20   28
        25   <u>35</u>
        30
        <u>35</u>

        3/5 becomes 21/35. Since the denominator needs to be multiplied by 7 to get to 35, the numerator needs to be multiplied by 7. 3 x 7 = 21 leading to 21/35

        5/7 becomes 25/35. Since the denominator needs to be multiplied by 5 to get to 35, the numerator needs to be multiplied by 5. 5 x 5 = 25 leading to 25/35.

21/35 can be compared to 25/35 since they have a common denominator.
21/35 is less than 25/35, which shows that 3/5 < 5/7
21/35 < 25/35
which means 3/5 < 5/7

2.   Cross-multiply to compare fractions. This way is quicker.

3/5 compared to 5/7

Cross-multiply 3 x 7 to get 21.
21 will represent the 3/5 side.
Cross-multiply 5 x 5 to get 25.
25 will represent the 5/7 side.

21 is less than 25.
This shows that 3/5 < 5/7

3.   Change fractions to decimals to compare.
To change a fraction into a decimal, divide the denominator into the numerator.

$3/5 = .600 = .6$ $5/7 = .714$

```
      .600                        .714
   5) 3.000                   7) 5.000
     - 30                       - 49
       00                         10
      -00                        - 7
       00                         30
```

.6 < .714

This shows that 3/5 < 5/7

Arithmetic Reasoning
Fractions - Pre-Test

1.   Change improper fraction to mixed number.

      $9/6 =$

2.   Change mixed number to improper fraction.

      $4 \ \ 2/6 =$

3. $4/9 + 1/3 =$

4. $5/6 - 1/4 =$

5. $2/8 \times 3/5 =$

6. $8/10 \div 2/5 =$

25

Arithmetic Reasoning
Fractions - Pre-Test with Answers

1. Change improper fraction to mixed number.

   $9/6 = 1 \ 3/6 = 1 \ 1/2$

2. Change mixed number to improper fraction.

   $4 \ 2/6 = 26/6$

3. $4/9 + 1/3 =$

   Find a common denominator.
   Add the numerators. Reduce if necessary.
   $4/9 + 3/9 = 7/9$

4. $5/6 - 1/4 =$

   Find a common denominator.
   Subtract the numerators. Reduce if necessary.
   $10/12 - 3/12 = 7/12$

5. $2/8 \times 3/5 =$

   Multiply the numerators.
   Multiply the denominators.
   Reduce if necessary.
   $2/8 \times 3/5 = 6/40 = 3/20$

6. $8/10 \div 2/5 =$

   Keep, change, flip.
   Multiply the numerators.
   Multiply the denominators.
   Reduce if necessary.
   $8/10 \times 5/2 = 40/20 = 2$

Mixed Numbers Review
Adding Mixed Numbers

3 1/4 + 5 2/3 =

Find a common denominator, then add the whole numbers and add the fractions.

List the multiples for 4 and 3.

4	3
8	6
12	9
	12

The least common denominator for 4 and 3 is 12.

Whatever you do to the denominator, you do to the numerator.

1/4 becomes 3/12 because 1 x 3 = 3
Since we multiplied the denominator by 3, we need to multiply the numerator by 3.
1 x 3 = 3 leading to 3/12

2/3 becomes 8/12 because 2 x 4 = 8
Since we multiplied the denominator by 4, we need to multiply the numerator by 4.
2 x 4 = 8 leading to 8/12

Add the fractions and add the whole numbers.

3 3/12 + 5 8/12 = 8 11/12

Subtracting Mixed Numbers

To subtract mixed numbers, first find a common denominator.

3 3/5 – 2 1/3 =

List the multiples of 5 and 3 to find the common denominator.

5	3
10	6
15	9
20	12
25	15

The least common denominator for 3 and 5 is 15.

Whatever you do to the denominator, you do to the numerator.

3 3/5 becomes 3 9/15 because 3 x 3 = 9. Since we multiplied the denominator by 3, we need to multiply the numerator by 3. 3 x 3 = 9 leading to 3 9/15.

2 1/3 becomes 2 5/15 because 1 x 5 = 5. Since we multiplied the denominator by 5, we need to multiply the numerator by 5. 1 x 5 = 5 leading to 2 5/15.

3 9/15 - 2 5/15 = 1 4/15

Subtract the fractions, then subtract the whole numbers.

In <u>some</u> cases of subtraction of mixed numbers, it becomes more complicated. Borrowing is needed in some cases. Here is an example that requires an additional step.

3 1/3 - 2 1/2 =

First find a common denominator.

List the multiples of 3 and 2.

3	2
<u>6</u>	4
9	<u>6</u>

The least common denominator for 3 and 2 is 6.

The additional step is required because 1/2 is bigger than 1/3 and cannot be taken away from 1/3. Borrowing is required. First, find a common denominator.

3 1/3 - 2 1/2 =

3 2/6 - 2 3/6 =

This is where there is an issue. You cannot subtract 3/6 from 2/6.
The next step is to borrow from the 3 in 3 2/6. Turn the 3 into a 2.
Since this problem has the fraction in sixths, the 1 that is borrowed becomes 6/6.
The 6/6 is then added to the 2/6.
3 2/6 becomes 2 2/6 + 6/6 = 2 8/6
Now we are able to subtract. Subtract the fractions. Subtract the whole numbers.
2 8/6 - 2 3/6 = 5/6

Another way to solve this problem is to change the mixed numbers into improper fractions.
3 1/3 - 2 1/2 =

10/3 – 5/2 =

Then find a common denominator.

List the multiples for 3 and 2.

3	2
6	4
9	6

The common denominator would be 6.

10/3 becomes 20/6 because 10 x 2 = 20. Since we multiplied the denominator by 2, we need to multiply the numerator by 2. 10 x 2 = 20 leading to 20/6.

5/2 becomes 15/6 because 5 x 3 = 15. Since we multiplied the denominator by 3, we need to multiply the numerator by 3. 5 x 3 = 15 leading to 15/6.

20/6 – 15/6 = 5/6

Multiplying Mixed Numbers

Change mixed numbers to improper fractions.
Multiply the top. Multiply the bottom. Reduce.
Turn the improper fraction back into a mixed number.

Do NOT find a common denominator.

3 3/4 x 5 1/2 =

15/4 x 11/2 = 165/8 = 20 5/8

Dividing Mixed Numbers

Change mixed numbers into improper fractions.
Keep, change, flip.
Keep the first fraction.
Change division to multiplication.
Flip the second fraction over.
Multiply the top. Multiply the bottom. Reduce.
Turn the improper fraction back into a mixed number.
Do NOT find a common denominator.

2 3/7 ÷ 1 5/8 =

17/7 ÷ 13/8 =

17/7 x 8/13 = 136/91 = 1 45/91

Multiplication of a Fraction by a Whole Number or Multiplication of Whole Number by Fraction

When presented with a whole number multiplied by a fraction or a fraction multiplied by a whole number, place the whole number over 1.
Multiply the numerators.
Multiply the denominators.
Reduce if necessary.

Fraction Multiplied by Whole Number
Example: 1/2 x 4 = 1/2 x 4/1 = 4/2 = 2

Whole Number Multiplied by Fraction
Example: 5 x 1/3 = 5/1 x 1/3 = 5/3 = 1 2/3

Division of a Fraction by a Whole Number or Division of a Whole Number by a Fraction

Put whole number over 1.
Keep first fraction, change division to multiplication, flip second fraction over.
Multiply the numerators.
Multiply the denominators.
Reduce if necessary.

Fraction Divided by a Whole Number
Example: 3/4 ÷ 5 =
              3/4 ÷ 5/1 =
              3/4 x 1/5 = 3/20

Whole Number Divided by a Fraction
Example: 5 ÷ 2/3 =
              5/1 ÷ 2/3 =
              5/1 x 3/2 = 15/2 = 7 1/2

Multiplication of a Mixed Number by a Whole Number or Multiplication of a Whole Number by a Mixed Number

Turn the whole number into a fraction by placing it over 1.
Change the mixed number into an improper fraction.
Multiply the numerators.
Multiply the denominators.
Reduce.

Multiplying a Mixed Number by a Whole Number

Example: 1 3/5 x 5 =
                    8/5 x 5/1 = 40/5 = 8

Multiplying a Whole Number by a Mixed Number

Example: 4 x 3 2/6 =
                    4/1 x 20/6 = 80/6 = 40/3 = 13 1/3

Division of a Mixed Number by a Whole Number or Division of a Whole Number by a Mixed Number

Turn the whole number into a fraction by placing it over 1.
Change the mixed number into an improper fraction.
Keep the first fraction.
Change division into multiplication.
Flip the second fraction over.
Multiply the numerators.
Multiply the denominators.
Reduce.

Dividing a Mixed Number by a Whole Number

Example: 4 2/7 ÷ 2 =
                    30/7 ÷ 2/1 =
                    30/7 x 1/2 = 30/14 = 15/7 = 2 1/7

Dividing a Whole Number by a Mixed Number

Example: 5 ÷ 3 2/6 =
                    5/1 ÷ 20/6 =
                    5/1 x 6/20 = 30/20 = 3/2 = 1 1/2

Mixed Numbers and Comparing Fractions
Pre-Test

1.   5 1/2
       + 3 1/6

2.   4 1/4
       - 2 3/5

3.   3 2/5 x 4 1/

4.   6 3/4 ÷ 1 1/2 =

5.   Put the following fractions in order from least to greatest.

       3/4 , 2/3 , 4/7, and 5/8

Julie A. Hyers ©2020

Mixed Numbers and Comparing Fractions
Pre-Test with Answers

1.   5 1/2   5 3/6
       + 3 1/6   + 3 1/6
                          8 4/6 = 8 2/3

2.   4 1/4   4   5/20   3   25/20
       - 2 3/5   - 2 12/20   - 2   12/20
                                          1   13/20

    12/20 cannot be subtracted from 5/20
    Borrow from the 4. Make the 4 into 3.
    The 1 that is borrowed becomes 20/20.
    Add 20/20 to 5/20 to become 25/20.
    Then subtract.

3.   3   2/5 x 4 1/3 =

    17/5 x 13/3 = 221/15 = 14 11/15

4.   6 3/4 ÷ 1 1/2 =

    27/4 ÷ 3/2 =

    27/4 x 2/3 = 54/12 = 4 6/12 = 4 1/2

5.   Put the following fractions in order from least to greatest.

    3/4, 2/3, 4/7, and 5/8

    Change to decimals to compare.

    3/4 = .7500
    2/3 = .6666
    4/7= .5714
    5/8 = .6250

    In order from least to greatest:
    4/7, 5/8, 2/3, 3/4

33

Percentage Review

Great Formula to Know

is/of = %/100

Example:
What percent of 80 is 60?

is/of = %/100

"is" is 60
"of" is 80
100 is always written under the percent.
In this case, the "percent" is missing.

60/80 = x/100
80x = 6000
x = 75%

It would also have been possible to reduce before cross-multiplying.
60/80 = x/100
Cross-multiply and solve for x.
80x = 6000
Reduce by removing one zero from each side of the equation.
8x = 600
x = 75%

Example:
What is 40% of 20?

is/of = %/100

"of" is 20
"percent" is 40%
100 is always written under the percent.
In this case, the "is" is missing.

x/20 = 40/100

Cross-multiply and solve for x.

100x = 800
x = 8

Example:
70 is 25% of what number?
is/of = %/100
"is" is 70
"percent" is 25%
100 is always written under the percent.
In this case, the "of" is missing.

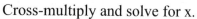

Cross-multiply and solve for x.
70/x = 25/100
7000 = 25x
280 = x

It is possible to reduce the 25/100 to 1/4.
70/x = 25/100
70/x = 1/4
280 = x

Percent Increase and Decrease Formula

difference/original = %/100

This formula is to used to find the percent increase or decrease.
The same formula is used for increase or decrease.
Subtract the 2 values on the top.
Write the original on the bottom.
Cross-multiply to solve.

difference/original = %/100

Example:
If a television costs $300 and is on sale for $240, what was the percent decrease?

The difference in price is 300-240 = 60
Original is 300.

The formula would be:
300-240/300 = %/100
60/300 = %/100

Cross-multiply to solve for the percentage.
6000 = 300x
20 = x

The television's price decreased by 20%.

The same formula is used for a percent increase question.

Example:
If a ring cost $110 and the price went up to $130, what was the percent increase?

difference/original = %/100

130 - 110/110 = x/100
20/110 = x /100

Cross-multiply and solve for x.
2000 = 110x
200 = 11x
18.18% = x

The ring's price increased by 18.18%

Multiple Percent Discounts

Some questions can ask about multiple percent discounts taken off an item.

Example: If a table costs $400 and is on sale for 20% off and then an additional 10% is taken off, what is the final sale price?

Do NOT add up the 20% and the 10% and think this is a 30% discount. That is a trap into which many people fall.

The percent reductions must be done one at a time.

Work on the first discount of 20% off.
Change the percent to a decimal.
20% = .20
Multiply 400 by .20 and subtract the result from 400.
400 x .20 = 80
400 − 80 = 320

Work on the second discount of 10% off.
Change the percent to a decimal.
10% = .10
Multiply 320 by .10 and subtract the result from 320.
320 x .10 = 32
320 − 32 = 288
The final answer is $288.

Diluting a Liquid

There is a type of question that involves diluting a liquid with water and figuring out what percentage of the original item is still in the liquid. It could be juice, antifreeze, paint, etc.

Example:
If there are 40 ounces of a drink that consists of 60% juice and 5 ounces of water is added to the drink, what is the percentage of juice in the new mixture?

Find how many ounces of juice are in the original amount.
Multiply 40 by 60%.
Change 60% to a decimal of .60
40 x .60 = 24
There are 24 ounces of juice in the drink.
Total liquid is 40 + 5 = 45 ounces
juice/total = %/100
24/45 = x/100
Cross-multiply and solve for x.
2400 = 45x
53.33% = x
The percentage of juice in the new mixture is 53.33%

Figuring Out Original Price When Sale Price Is Given

There is a type of question where the sale price is given.
It mentions the percent discount off the original price, and it asks what the original price was.

Example:
The sale price of a couch was $1200. It was 25% off the original price. What was the original price?

A trap many people fall into is finding 25% of the $1200. This is not the way to find it.
25% was taken off the original price.
Since 25% was taken off the original price, this shows that there is 75% left.
Therefore $1200 is 75% of the original price.
A proportion could be set up.

Sale price/original price = %/100
We do not know the original price so that will be the x.
1200/x = 75/100
Cross-multiply and solve for x.
120000 = 75x
1600 = x
The original price was $1600.

It would be possible to reduce this example before cross-multiplying.
1200/x = 75/100

75/100 could be reduced to 3/4.
1200/x = 3/4
4800 = 3x
1600 = x
The original price was $1600.

Questions About Percent of a Whole

These types of questions are set up like the proportions of is/of = %/100.

Example:
In a group of 80 children, there are 70 boys, what percent are boys?

boys/total children = %/100
70/80 = x/100
Reduce 70/80 to become 7/8.
7/8 = x/100
8x = 700
x = 87.5%

Questions About Population Increase and Decrease

When there is a question about population increase or decrease and no original population size is given, use 100 as the starting population.

Assume the starting population is 100.
It is easier to solve if you assume the starting population is 100.

Example:
If the population increased 20% from 1970 to 1980, then it decreased 5% from 1980 to 1990, what was the percent increase from 1970 to 1990?

Start with 100 as the starting population.
If the population increased by 20% from 1970 to 1980, then it is 120 in 1980.
If the population decreased by 5% from 1980 to 1990, find 5% of 120.

120 x .05 = 6

120 − 6 = 114

The original population was 100, and in the end, it is 114, which is a 14% increase.

Arithmetic Reasoning
Percentage - Pre-Test

1.   What percent of 80 is 50?

2.   What is 30% of 120?

3.   15 is 20% of what number?

4.   A window sells for $80.00 and is on sale for $60.00, what is the percent discount?

5.   If a motorcycle costs $500.00 and the price increases to $600.00, what is the percent increase?

6. If a leather jacket costs $200.00 and is on sale for 30% off and then an additional 15% is taken off, what is the final sale price?

7. If 80 ounces of a drink consists of 20% juice and 10 ounces of water are added to the drink, what is the percentage of juice in the new mixture?

8. The sale price of a dining room set is $1600.00, which is 20% off the original price. What is the original price?

9. In a group of 45 people, 25 are female, what percent are female?

10. If a town's population increased by 30% from 1990 to 2000 and decreased by 20% from 2000 to 2010, what was the percent increase in population from 1990 to 2010?

Arithmetic Reasoning
Percentage - Pre-Test Answers

1.   What percent of 80 is 50?

  is/of = %/100

  $50/80 = x/100$
  $5000 = 80x$
  $500 = 8x$
  $62.5\% = x$

2.   What is 30% of 120?

  is/of = %/100

  $x/120 = 30/100$
  $100x - 3600$
  $x = 36$

3.   15 is 20% of what number?

  is/of = %/100

  $15/x = 20/100$
  $1500 = 20x$
  $150 = 2x$
  $75 = x$

4.   A window sells for $80.00 and is on sale for $60.00, what is the percent discount?

  Percent increase and decrease formula
  difference/original = %/100

  $80 - 60/80 = x /100$
  $20/80 = x/100$
  $2/8 = x/100$
  $1/4 = x/100$
  $100 = 4x$
  25% discount

41

5. If a motorcycle costs $500.00 and the price increases to $600.00, what is the percent increase?

Percent increase and decrease formula
difference/original = %/100
600-500/500 = x/100
100/500 = x/100
1/5 = x/100
5x = 100
x = 20% increase

6. If a leather jacket costs $200.00 and is on sale for 30% off and then an additional 15% is taken off, what is the final sale price?

200 x .30 = 60
200 – 60 = 140
140 x .15 = 21
140 – 21 = $119

7. If 80 ounces of a drink consists of 20% juice and 10 ounces of water are added to the drink, what is the percentage of juice in the new mixture?

Find how many ounces of juice are in the original amount.
80 x .20 = 16 ounces
Total liquid = 80 +10 = 90
juice/total = %/100
16/90 = x/100
1600 = 90x
160 = 9x
17.77% = x

8. The sale price of a dining room set is $1600.00, which is 20% off the original price. What is the original price?

$1600 is 80% of the original price
1600/x = 80/100
160000 = 80x
16000 = 8x
$2000 = x

9.   In a group of 45 people, 25 are female, what percent are female?

Female/total = %/100
25/45 = %/100
25/45 = x/100
Reduce 25/45 to become 5/9
5/9 = x/100
500 = 9x
55.55% = x

10.   If a town's population increased by 30% from 1990 to 2000 and decreased by 20% from 2000 to 2010, what was the percent increase in population from 1990 to 2010?

Assume the starting population is 100.
If the population increases 30% from 1990 to 2000, then it is 130 in 2000.
If the population decreased by 20% from 2000 to 2010, find 20% of 130
130 x .20 = 26
130 − 26 = 104
The original population was 100, and in the end, it is 104, which is a
4% increase.

Percent, Fraction, Decimal Conversion Review

Changing a Fraction to a Decimal

Divide the denominator into the numerator.
Add a decimal point and a few zeros after the numerator, then divide.
Example:
$11/9 = 1.2222$

Changing a Mixed Number to a Decimal

Keep the whole number as a whole number.
Divide the denominator into the numerator, add a decimal and a few zeros to the numerator.
Example:
$3 \quad 1/5 = 3.20 = 3.2$

Changing a Fraction to a Percent

Write it as a proportion. Write the fraction and set it equal to x/100.
Cross-multiply and solve for x.
Example:
$6/4 = x/100$
$600 = 4x$
$150\% = x$

Changing a Mixed Number to a Percent

Change the mixed number into an improper fraction.
Write it as a proportion.
Write the fraction and set it equal to x/100.
Cross-multiply and solve for x.
Example:
$4 \quad 1/4 = 17/4$
$17/4 = x/100$
$1700 = 4x$
$425\% = x$

Changing a Decimal to a Fraction

Place the number over 10, 100, 1000, etc., depending on how many numbers appear after the decimal point.
If there is one digit after the decimal point, place the number over 10.

If there are two digits after the decimal point, place the number over 100.
If there are three digits after the decimal point, place the number over 1000, and so on.
Reduce if necessary.
Examples:
.5 = 5/10 = 1/2
.75 = 75/100 = 3/4
.278 = 278/1,000 = 139/500

If the decimal includes a whole number, keep the whole number as a whole number and make the decimal part into a fraction.
Example:
3.25 = 3 25/100 = 3 1/4

Change a Decimal to a Percent

D→P
Move the decimal point 2 places to the right.
Examples:
.12 = 12%
.07 = 7%
1.35 = 135%

Changing a Percent to a Decimal

D ←P
Move the decimal point 2 places to the left.
Examples:
75% = .75
3% = .03
125% = 1.25

Changing a Percent to a Fraction

Place the number over 100 and reduce.
Examples:
5% = 5/100 = 1/20
150% = 150/100 = 1.50 = 1 1/2
80% = 80/100 = 4/5
10.25% = 10.25/100 = 1025/10,000 = 205/2,000 = 41/400

Common Fraction, Decimal, Percent Conversions

Fraction	Decimal	Percent
1/4	.25	25%
2/4 = 1/2	.50	50%
3/4	.75	75%
4/4	1.00	100%
1/5	.20	20%
2/5	.40	40%
3/5	.60	60%
4/5	.80	80%
5/5	1.00	100%
1/3	.3333	33.33%
2/3	.6666	66.66%
3/3	1.0000	100%
1/10	.10	10%
2/10 = 1/5	.20	20%
3/10	.30	30%
4/10 = 2/5	.40	40%
5/10 = 1/2	.50	50%
6/10 = 3/5	.60	60%
7/10	.70	70%
8/10 = 4/5	.80	80%
9/10	.90	90%
10/10	1.00	100%

Arithmetic Reasoning
Percent, Decimal, Fraction Conversions - Pre-Test

1.   Change fraction to decimal.
       3/7 =

2.   Change mixed number to decimal.
       1 8/9 =

3.   Change fraction to percent.
       5/8 =

4.   Change a mixed number to percent.
       4   1/6 =

5.   Change decimal to fraction.
       .45 =

6.   Change decimal to fraction.
       3.56 =

7.   Change decimal to percent.
      .72 =

8.   Change decimal to percent.
      4.76 =

9.   Change percent to decimal.
      29% =

10.   Change percent to decimal.
      356% =

11.   Change percent to fraction.
      82% =

12.   Change percent to fraction.
      965% =

Arithmetic Reasoning
Percent, Decimal, Fraction Conversions - Pre-Test with Answers

1. Change fraction to decimal.

    $3/7 = .428$

    Divide the denominator into the numerator.
    $3.00 \div 7 = .428$

2. Change mixed number to decimal.

    $1 \ 8/9 = 1.88$

    Keep the whole number as 1.
    Divide the denominator into the numerator.
    $8.00 \div 9 = .88$
    The answer is 1.88

3. Change fraction to percent.

    $5/8 = .625$

    $5/8 = x/100$
    Cross-multiply and solve for x.
    $500 = 8x$
    $62.5\% = x$

4. Change a mixed number to percent.

    $4 \ 1/6 = 25/6 = 4.1666 = 416.66\%$

    Change the mixed number into an improper fraction.
    Set up a proportion.
    $25/6 = x/100$
    Cross-multiply and solve for x.
    $2500 = 6x$
    $416.66\% = x$

5.   Change decimal to fraction.

.45 = 45/100 = 9/20

Check to see how many digits are after the decimal point.
In this case, there are 2 digits after the decimal point.
That means 45 goes over 100 since 100 has 2 zeros.
.45 = 45/100
Reduce 45/100 to 9/20.

6.   Change decimal to fraction.

3.56 = 3 56/100 = 3 28/50 = 3 14/25

Keep 3 as a whole number.
Check to see how many digits are after the decimal point.
In this case there are 2 digits after the decimal point.
That means 56 goes over 100 since 100 has 2 zeros.
3.56 = 3 56/100 and reduce to 3 28/50 and then reduce to 3 14/25.

7.   Change decimal to percent.

.72 = 72%

Move the decimal point 2 places to the right.

8.   Change decimal to percent.

4.76 = 476%

Move the decimal point 2 places to the right.

9.   Change percent to decimal.

29% = .29

Move the decimal point 2 places to the left.

10. Change percent to decimal.

356% = 3.65

Move the decimal point 2 places to the left.

11. Change percent to fraction.

82% = 82/100 = 41/50

Place the number over 100 and reduce.

12. Change percent to fraction.

965% = 965/100 = 9 65/100 = 9   13/20

Place the number over 100 and reduce.

Proportion Review

Proportion is the relationship between two numbers expressed in fraction form.

6/10 = x/5
Cross-multiply to solve for x.
30 = 10x
Divide by 10 on each side to solve for x.
3 = x

Shadow and Flagpole Question

This question is a <u>must know</u> for the ASVAB.

If a tree is 20 ft tall and its shadow is 8 ft tall and a man is 6 ft tall, how tall is the man's shadow?

object/shadow = object/shadow
20/8 = 6/x
Cross-multiply and solve for x.
20x = 48
x = 48/20 = 2 8/20 = 2 2/5 ft. = 2.4 ft.

Inches on a Map Question

This is another common question on the ASVAB.

On a map, if 1 inch represents 45 miles, how many miles does 3 1/4 inches represent?

To make it easier, change 3 1/4 into a decimal 3.25.

inches/miles = inches/miles
1/45 = 3.25/x
Cross-multiply and solve for x.
x = 45 x 3.25 = 146.25 miles or 146 1/4 miles

Recipe Question

Proportions can be used to solve certain recipe questions.

If an apple pie for 12 people has 2 1/4 cups of apples, how many cups of apples are needed for an apple pie for 16 people?

people/cups = people/cups

Change 2 1/4 into a decimal, 2.25, to make it easier to solve.
12/2.25 = 16/x
Cross-multiply and solve for x.
12x = 36
x = 3 cups

Ounces and Cost Questions

When there are questions about the weight of a product and the price, a proportion is useful to solve it. To solve this problem, remember that there are 16 ounces in a pound.

Change pounds to ounces and set up a proportion of ounces over cost.

If 2 lb 4 oz of chopped liver costs $12.00, how much would 6 oz of chopped liver cost?

To solve this problem, you need to know that there are 16 ounces in 1 pound.

1 lb = 16 ounces
2 lb 4 oz = (2 x 16) + 4 = 32 + 4 = 36 oz

oz/cost = oz/cost
36/12 = 6/x
Reduce 36/12 to 3/1 by dividing by 3.
3/1= 6/x
Cross-multiply and solve for x.
3x = 6
x = $2.00

If 8 oz of grapes costs $1.40, how much would 4 lb of grapes cost?

ounces/cost = ounces/cost

To solve this problem, you need to know there are 16 ounces in 1 pound.

Change 4 lb to ounces.
There are 16 ounces in a pound. 4 lb is 64 ounces.
8/1.40 = 64/x
Reduce 8 and 64 to become 1 and 8 by dividing by 8.
Cross-multiply to solve for x.
1/1.40 = 8/x
x = $11.20

Julie A. Hyers ©2020

Arithmetic Reasoning
Proportions - Pre-Test

1. If a tree is 15 ft tall and its shadow is 9 ft tall and a flagpole is 12 ft tall, how tall is the flagpole's shadow?

2. On a map, if 1 inch represents 15 miles, how many miles does 2 1/5 inches represent?

3. If blueberry pie for 10 people has 3 1/4 cups of blueberries, how many cups of blueberries are needed for a blueberry pie for 15 people?

4. If 1 lb 4 oz of fish costs $16.00, how much would 8 oz of fish cost?

5. If 4 oz of candy costs $0.80, how much would 3 lb of candy cost?

Arithmetic Reasoning
Proportions - Pre-Test with Answers

1. If a tree is 15 ft tall and its shadow is 9 ft tall and a flagpole is 12 ft tall, how tall is the flagpole's shadow?

    15/9 = 12/x
    Cross-multiply and solve.
    15x = 108
    x = 108/15 = 7 3/15 ft = 7 1/5 ft = 7.2 ft

2. On a map, if 1 inch represents 15 miles, how many miles does 2 1/5 inches represent?

    To make it easier, change 2 1/5 into a decimal, 2.2
    inches/miles = inches/miles
    1/15 = 2.2/x
    x = 15 x 2.2 = 33 miles

3. If blueberry pie for 10 people has 3 1/4 cups of blueberries, how many cups of blueberries are needed for a blueberry pie for 15 people?

    people/cups = people/cups
    10/3.25 = 15/x
    10x = 48.75
    x = 4.875 cups

    This could have been reduced since both 10 and 15 can be divided by 5.
    2/3.25 = 3/x
    2x = 9.75
    x = 4.875 cups

4. If 1 lb 4 oz of fish costs $16.00, how much would 8 oz of fish cost?

    1lb 4 oz = 20 oz
    oz./cost = oz./cost
    20/16.00 = 8/x
    Reduce 20/16.00 to 5/4.00
    5/4.00 = 8/x
    5x = 32.00
    x = $6.40

5.   If 4 oz of candy costs $0.80, how much would 3 lb of candy cost?

ounces/cost = ounces/cost
Change 3 lb to ounces. There are 16 ounces in a pound, so 3 lb is 48 ounces.
4/.80 = 48/x
Reduce 4 and 48 to become 1 and 12.
1/.80 = 12/x
x = $9.60

Ratio Review

Ratio is the relationship between 2 things or numbers.

Example:
Boy to girl ratio is 5:3.
Can be written in three ways: 5:3, 5/3, or 5 to 3.
That means there are 5 boys for every 3 girls.
It could be 10 boys and 6 girls, 50 boys and 30 girls, and so on.

Ratio Questions

Example:
If there is a 7:4 ratio of cats to dogs, and there are 35 cats, how many dogs are there?

cats/dogs = cats/dogs

7/4 = 35/x
Reduce 7 and 35 to become 1 and 5.
Reducing left to right or up and down is possible in proportions.
Cross-multiply and solve for x.
1/4 = 5/x
x = 20 dogs

Ratio Involving a Sum

Example:
If a woman donates \$125,000 to charity and she gives her money to a school and a charity in a 4:1 ratio, how much money did she give to the school ?

Most ratio problems are solved as proportions, but this one is not. This is a different case where the amount of money is the total sum of money. Addition of the ratios is required.

school: charity donation ratio is 4:1

4x + x = 125,000
5x = 125,000
x = 25,000
Usually when we get to x, the problem is solved.
However, in this case, the value of x is not the answer.
The value of the school donation is 4x.
The school donation is 4x = 4(25,000) = 100,000
The donation to the school is \$100,000.

Example:
If there are 20 women and 15 men working in an office, what is the ratio of women to people in the office?

Women/total people
Women + men = 20 + 15 = 35
Women/total people = 20/35 = 4/7

Ratio Involving Dimensions

Example:
If the dimensions of a kitchen are 15 ft by 20 ft and the dimension of a living room are 30 ft by 40 ft, what is the ratio of the area of the kitchen to the area of the living room?

	L x W
Kitchen	15ft x 20ft
	2 x 2 = 4
Living Room	30ft x 40ft

Divide each dimension of the kitchen into the corresponding dimension of the living room. Multiply the results to get 4.

The length of the living room is 2 times longer than the length of the kitchen.
The width of the living room is 2 times longer than the width of the kitchen.

The ratio of the kitchen to the living room is 1:4, since the living room is 4 times bigger than the kitchen.

Example:
How many boxes of crackers measuring 2 inches long by 3 inches wide by 4 inches deep will fit into a box measuring 12 inches long by 15 inches wide by 16 inches deep?

L   x   W   x   D
2 in. x   3 in. x   4 in.
6   x 5   x 4   = 120 boxes
12 in. x 15 in. x 16 in.

Divide each dimension of the smaller box into the corresponding dimension of the larger box. Since all the dimensions are measured in inches, no conversions are needed.

Multiply the results to get the answer of 120 boxes.

Example:
If a tile is 6 inches by 6 inches, how many tiles are needed in a room measuring 30 ft by 40 ft?

To solve this problem, you need to know there are 12 inches in 1 foot.

Each tile is 6 in. x 6 in.
The room is 30 ft x 40 ft
Change ft into in.
30 ft = 360 inches
40 ft = 480 inches

L   x   W
6 in. x 6 in.
60   x   80   =   4,800 tiles
360 in. x 480 in.

The tiles are measured in inches while the room is measured in feet.

Convert the room into inches.
Then divide the dimensions of the tile into the corresponding dimensions of the room.
Multiply the results to find that the room requires a total of 4,800 tiles.

Arithmetic Reasoning
Ratios - Pre-Test

1.   If there is an 8:3 ratio of cats to dogs, and there are 24 cats, how many dogs are there?

2.   If a woman donates $75,000 to charity and she gives her money to a hospital and a museum in a 2:1 ratio, how much money did she give to the hospital?

3.   If there are 10 women and 8 men working in an office, what is the ratio of women to people in the office?

4.   If the dimensions of a living room are 10 ft by 12 ft and the dimensions of a basement are 90 ft by 60 ft, what is the ratio of the area of the living room to the area of the basement?

5. How many boxes of pretzels measuring 3 inches long by 4 inches wide by 5 inches deep will fit into a box measuring 15 inches long by 16 inches wide by 20 inches deep?

6. If a tile is 10 inches by 10 inches, how many tiles are needed in a room measuring 10 ft by 20 ft?

5. How many boxes of pretzels measuring 3 inches long by 4 inches wide by 5 inches deep will fit into a box measuring 15 inches long by 16 inches wide by 20 inches deep?

6. If a tile is 10 inches by 10 inches, how many tiles are needed in a room measuring 10 ft by 20 ft?

Arithmetic Reasoning
Ratios - Pre-Test with Answers

1. If there is an 8:3 ratio of cats to dogs, and there are 24 cats, how many dogs are there?

    cats/dogs = cats/dogs
    8/3 = 24/x
    Reduce 8 and 24 to become 1 and 3.
    Reducing left to right or up and down is possible in proportions.
    1/3 = 3/x
    x = 9 dogs

2. If a woman donates $75,000 to charity and she gives her money to a hospital and a museum in a 2:1 ratio, how much money did she give to the hospital?

    hospital: museum donation ratio is 2:1
    2x + x = 75,000
    3x = 75,000
    x = 25,000
    Hospital donation is 2x = 2(25,000) = 50,000
    The donation to the hospital is $50,000.

3. If there are 10 women and 8 men working in an office, what is the ratio of women to people in the office?

    women/total people
    women + men = 10 + 8 = 18
    women/total people = 10/18 = 5/9

4. If the dimensions of a living room are 10 ft by 12 ft and the dimensions of a basement are 90 ft by 60 ft, what is the ratio of the area of the living room to the area of the basement?

    Living room   10 ft x 12 ft

                   9   x   5   =   45

    Basement   90 ft x 60 ft

The ratio of the living room to the basement is 1:45 since the basement is 45 times bigger than the living room.

5.   How many boxes of pretzels measuring 3 inches long by 4 inches wide by 5 inches deep will fit into box measuring 15 inches long by 16 inches wide by 20 inches deep?

3 in. x 4 in. x 5 in.
5   x 4   x 4   = 80 boxes
15 in. x 16 in. x 20 in.

Since the small pretzel box has dimensions of 3 inches x 4 inches x 5 inches, divide each of these dimensions into the dimensions of the large box, which is 15 inches x 16 inches x 20 inches.
The result is 5 x 4 x 4 which needs to be multiplied together to get 80 boxes.
The total number of boxes that fit into the large box is 80 boxes.

6.   If a tile is 10 inches by 10 inches, how many tiles are needed in a room measuring 10 ft by 20 ft?

Each tile is 10 in. x 10 in.
The room is 10 ft x 20 ft
Change ft into in.
10 ft = 120 inches
20 ft = 240 inches

10 in. x   10 in.
12 x   24   = 288 tiles
120 inches x 240 inches

In this question, the tiles are measured in inches, and the room is measured in feet. The dimensions of the room need to be converted into inches to solve the problem.
To solve this problem, you need to remember that there are 12 inches in a foot.
Each tile measures 10 in. by 10 in. The room measures 10 ft x 20 ft.
10 ft = 120 in. and 20 ft = 240 in.
Divide the dimensions of the tile into the dimensions of the room.
10 in. x 10 in. is divided into 120 in. x 240 in.
The result is 12 x 24 which needs to be multiplied to give the result of 288 tiles.

Tax, Tip, Commission, and Overtime Review

Sales tax, tip, and commission questions involve percentages.
To solve these problems, change the percent to a decimal and multiply.

Sales Tax

Change the percent to a decimal and multiply by total price to find the sales tax.

If sales tax is 8.5%, how much is the sales tax on a pair of shorts that costs $45.00?

To find sales tax, change the percent to a decimal. Then multiply.
8.5% = .085
45 x .085 = 3.825 = $3.83

Tip

Change the percent to a decimal and multiply by total price to find the tip.

Allison dined at a restaurant. Her bill came to $25.40, and she wants to give a 15% tip. How much would she leave for the tip?

To find tip, change the percent to a decimal. Then multiply.
15% = .15
25.40 x .15 = $3.81

Commission

Commission questions have three steps.
1. There is base pay which involves multiplying the hourly rate by the number of hours worked.
2. There is commission, which involves multiplying the amount sold by the percent commission. Change the percent to a decimal and multiply.
3. Add base pay and commission to find total salary.

A salesperson earns $11.00 an hour and 14% commission on any purchases made. If he sells $3000 in merchandise and works for 40 hours, what was his total earned for the week?

Find the base pay by multiplying hourly rate by the hours worked.
40 x 11 = $440

Find the commission by changing the percent to a decimal and multiplying the percent by how much merchandise was sold.
Change the percent to a decimal. Then multiply.
.14% = .14
3000 x .14 = $420
Add the base pay and the commission.
$440.00 + $420.00 = $860.00

Overtime

To solve overtime questions, there are multiple steps.
The example below shows a quicker way to solve it than the traditional way of solving an overtime question.

Violet earns $15.50 an hour for 40 hours and earns 1 1/2 times her salary for overtime. If she works a total of 60 hours in a week, how much did she earn that week?

Subtract 40 hours from the total hours worked.
60 - 40 = 20
Multiply 20 by 1.5 = 30 hours.
Add 40 + 30 = 70 hours.
Multiply the hourly rate by 70.
15.50 x 70 = $1085.00

The more traditional way to solve overtime problems is done in the following way:

$15.50 x 40 = $620.00

Then multiply the hourly rate by 1.5 to figure out the overtime rate.
$15.50 x 1.5 = $23.25

Then subtract 40 from 60 to figure out how many overtime hours are worked.
Multiply the overtime hours by the overtime rate.
$23.25 x 20 = $465.00

Add the base pay to the overtime pay to find the total pay.

$620.00 + $465.00 = $1085.00

Arithmetic Reasoning
Tax, Tip, Commission, Overtime - Pre-Test

1.   If sales tax is 8.5%, how much is the sales tax on a suitcase that costs $75.00?

2.   Nicky dined at a restaurant. His bill came to $45.70, and he wants to give a 15% tip. How much would he leave for the tip?

3.   A salesperson earns $9.00 an hour and 11% commission on any sales made. If he sells $2500 in merchandise and works for 45 hours, what was his total earned for the week?

4.   Ann earns $12.50 an hour for 40 hours and earns 1 1/2 times her salary for overtime. If she works a total of 50 hours in a week, how much will she earn that week?

Arithmetic Reasoning
Tax, Tip, Commission, Overtime - Pre-Test with Answers

1.   If sales tax is 8.5%, how much is the sales tax on a suitcase that costs $75.00?

      75 x .085 = 6.375 = $6.38

2.   Nicky dined at a restaurant. His bill came to $45.70, and he wants to give a 15% tip. How much would he leave for the tip?

      45.70 x .15 = $6.855 = $6.86

3.   A salesperson earns $9.00 an hour and 11% commission on any sales made. If he sells $2500 in merchandise and works for 45 hours, what was his total earned for the week?

      Find the base pay by multiplying hourly rate by the hours worked.
      45 x 9 = $405
      Find the commission by changing the percent to a decimal and multiplying the percent by how much merchandise was sold.
      2500 x .11 = $275
      Add the base pay and the commission.
      $405.00 + $275.00 = $680.00

4.   Ann earns $12.50 an hour for 40 hours and earns 1 1/2 times her salary for overtime. If she works a total of 50 hours in a week, how much will she earn that week?

      Subtract 40 hours from the total hours worked.
      50 - 40 = 10
      Multiply 10 by 1.5 = 15 hours
      Add 40 + 15 = 55 hours
      Multiply the hourly rate by 55.
      12.50 x 55 = $687.50

      The more traditional way to solve overtime problems is done in the following way:
      $12.50 x 40 = $500.00
      Then multiply the hourly rate by 1.5 to figure out the overtime rate.
      $12.50 x 1.5 = $18.75
      Then subtract 40 from 50 to figure out how many overtime hours are worked.
      Multiply the overtime hours by the overtime rate.
      $18.75 x 10 = $187.50
      Add the base pay to the overtime pay to find the total pay.
      $500.00 + $187.50 = $687.50

Interest Formula Review

Interest formula

I = p x r x t

Interest = principal x rate x time

Principal is how much money was put into the bank.
Rate is the interest rate, which is written as a percentage.
Before multiplying, it needs to be changed into a decimal.
Move the decimal point 2 places to the left to do this.
For example, 5% becomes .05
Time is the number of years the money is invested.
For 1 year, t =1
For 6 months, t = .5 because 6 months is 1/2 of a year.

When using the interest formula, you need to see what you are solving for.
You might need to find the interest, the principal, the rate or the time.
Read carefully to see which one is missing and solve accordingly.

Here is an example where the interest is missing, and you need to solve for interest.

If $1000 is invested at 5% for 2 years, how much interest did it earn?
I = p x r x t
Principal = $1000
Rate = 5% = .05
Time = 2

In this case, the interest is missing.
Interest is what you need to solve for.
Interest = ?
I = ?
Interest = 1000 x .05 x 2
Interest = 2000 x .05
Interest = $100

Here is an example where the principal is missing, and you need to solve for principal.

Sandra puts money in the bank at a rate of 3% for 4 years and earns $48.00 in interest.
How much money does she put in the bank?

Interest = principal x rate x time
Change 3% to .03

Interest = $48.00
Rate = 3% = .03
Time = 4 years

In this case, principal is missing.
Principal is what you need to solve for.
Principal = ?
P = ?

48 = p x .03 x 4
48 = p x .12
48 = .12p
Divide by .12
48 ÷ .12 =
4800 ÷ 12 = $400
Principal = $400

Here is an example where the interest rate is missing, and you need to solve for interest rate.

Samuel puts $300 in the bank for 6 years and earns $90 in interest. What was the interest rate?

Interest = principal x rate x time
Principal = $300
Time = 6 years
Interest = $90

In this case, the rate is missing.
Rate is what you need to solve for.
Rate = ?
R = ?

90 = 300 x r x 6
90 = 1800r
90 ÷ 1800 =
90.00 ÷ 1800 = .05 = 5%
Rate = 5%

Here is an example where the time is missing, and you need to solve for time.

Terry puts $1200 in the bank at a rate of 4% and earns $96 in interest. How many years did it take to earn that much interest?

Interest = principal x rate x time
Principal = $1200

Rate = 4%
Interest = $96

In this case the time is missing, and you need to solve for time.
Time = ?
T = ?

96 = 1200 x .04 x t
96 = 48t
2 = t
2 years = time

Compound Interest

Another type of interest question is compound interest.

Compound interest is different from simple interest. The previous examples were simple interest. Compound interest involves finding the interest for each year and adding it on before finding the interest for the next year.

Example:
Angelina put $600 in the bank at 4% for 3 years. The interest is compounded annually. Including interest, how much money will be in his bank account at the end of the 3 years?

Principal = $600
Rate = 4% = .04
Time = 3 years
Interest is what we need to solve for.

Since this is a compound interest question, we do not use 3 for t (time). We use 1 and solve it one year at a time, adding the interest on for each year before solving for the next year.

Year 1
Interest = principal x rate x time
I = 600 x .04 x 1
I = 24
At the end of year one, there is $24 in interest, which must be added onto the original $600, giving us $624.

$600.00 + $24.00 = $624.00

Year 2
We start with the $624.
I = p x r x t
I = 624 x .04 x 1 = 24.96
I = 24.96
At the end of year 2, there is $24.96 in interest, which must be added onto the $624.00, giving us $648.96.

$624.00 + 24.96 = $648.96

Year 3
We start with $648.96
I = 648.96 x .04 x 1 = 25.9584 = 25.96
 I = 25.96
At the end of year 3, there is $25.96 in interest, which must be added onto the $648.96

$648.96 + 25.96 = $674.92

As you can see, finding compound interest is very time consuming.
If the answer choices are not that close, you can use the simple interest formula to estimate the answer. If you use the simple interest formula for an estimate, the real answer to a compound interest question will be a little bit larger.

To solve this problem using a simple interest formula, use 3 years for time.
I = p x r x t
I = 600 x .04 x 3
I = 1800 x .04 = 72
Add $600.00 + $72.00 = $672.00
The answer of $672.00 is an estimate.

The real answer is $674.92 so you can see that the estimate of $672.00 is close to the real answer.

Depreciation

Depreciation questions talk about the value of an item reducing by a certain percentage over a period of time.

The percentage needs to be calculated one year at a time.

Example:
If Lexi's car is worth $11,000 and depreciates by 6% per year, how much will her car be worth in 2 years?
The percentage needs to be calculated one year at a time.
Figure out the percentage of the starting price.
Then subtract the amount from the starting price.

Year 1
$11,000 x .06 = $660
$11,000.00 – $660.00 = $10,340.00

The percentage needs to be calculated for the second year.

Year 2
$10,340 x .06 = $620.40
$10,340.00 - $620.40 = $9,719.60

The final value of the car after two years of depreciation is $9,719.60.

Arithmetic Reasoning
Interest Rate - Pre-Test

1.   Phoebe put $700.00 in a bank account that earns 3% interest. How much interest will she earn in 5 years?

2.   Stephanie puts money in the bank at a rate of 4% for 2 years and earns $36.00 in interest. How much money did she put in the bank?

3.   Donald puts $700 in the bank for 3 years and earns $63 in interest. What was the interest rate?

4.   Paul puts $1500 in the bank at a rate of 2% and earns $180 in interest. How many years did it take to earn that much interest?

5.   Vernon put $700 in the bank at 3% for 3 years. The interest is compounded annually. Including interest, how much money will be in his bank account at the end of the 3 years?

6.   If Lucas' car is worth $12,000 and depreciates by 8% per year, how much will his car be worth in 2 years?

Arithmetic Reasoning
Interest Rate - Pre-Test with Answers

1. Phoebe put $700.00 in a bank account that earns 3% interest. How much interest will she earn in 5 years?

   Interest = principal x rate x time
   Change 3% to a decimal .03.
   Principal = $700
   Rate = 3% = .03
   Time = 5 years
   Interest is missing.
   I = 700 x .03 x 5
   Multiply 700 x 5 first and then multiply by .03 (because we want to wait until the last step to multiply the decimal to make the steps easier.)
   3500 x .03 = $105.00

2. Stephanie puts money in the bank at a rate of 4% for 2 years and earns $36.00 in interest. How much money does she put in the bank?

   Interest = principal x rate x time
   Change 4% to .04
   Interest = $36.00
   Rate = 4% = .04
   Time = 2 years
   Principal is missing.
   36 = p x .04 x 2
   36 = p x .08
   36 = .08p
   Divide by .08
   36 ÷ .08 = $450

3. Donald puts $700 in the bank for 3 years and earns $63 in interest. What was the interest rate?

   Interest = principal x rate x time
   Principal = $700
   Time = 3 years
   Interest = $63
   Rate is missing.
   63 = 700 x r x 3
   63 = 2100r
   63 ÷ 2100 =
   63.00 ÷ 2100 = .03 = 3%

75

4.	Paul puts $1500 in the bank at a rate of 2% and earns $180 in interest. How many years did it take to earn that much interest?

Interest = principal x rate x time
Principal = $1500
Rate = 2%
Interest = $180
Time is missing.

180 = 1500 x .02 x t
180 = 30t
6 = t
6 years = time

5.	Vernon put $700 in the bank at 3% for 3 years. The interest is compounded annually. Including interest, how much money will be in his bank account at the end of the 3 years?

Compound interest

Compound interest needs to be figured out one year at a time. The interest of each year is added on and the interest formula is completed. The formula is completed one time for each year.

Interest = principal x rate x time
Principal = $700
Rate = 3%
Interest = $21

Year 1
I = p x r x t
Interest = principal x rate x time
I = 700 x .03 x 1
I = $21
$700.00 + $21.00 = $721.00

Year 2
I = p x r x t
Interest = principal x rate x time
I = 721 x .03 x 1
I = 21.63
721.00 + 21.63 = 742.63
$742.63

Year 3
I = p x r x t
Interest = principal x rate x time
I = 742.63 x .03 x 1
I = 22.278 = 22.28
742.63 + 22.28 = 764.91

$764.91

To estimate compound interest, use the simple interest formula.
I = p x r x t
I = 700 x .03 x 3
I = 2100 x .03 = 63
Add it to the principal
700.00 + 63.00 = $763.00

$763.00 is an estimate.
The real answer is a bit bigger than this. If the answer choices on the test are not close on the test, you can try to use the estimate instead of figuring out the compound interest. Select the choice that is a little bigger than the answer you picked. The estimate is $763.00. The real answer is $764.91 so you can see the estimate is close to the real answer.

6. If Lucas' car is worth $12,000 and depreciates by 8% per year, how much will his car be worth in 2 years?

    Year 1
    $12,000 x .08 = $960
    $12,000.00 – $960.00 = $11,040.00

    Year 2
    $11,040 x .08 = $883.20
    $11,040.00 - $883.20 = $10,156.80

Patterns and Sequences Review

When reviewing patterns and sequences, look at each number in the sequence.
Examine what happens between each number.

If the numbers go up, it is either addition or multiplication.
Check the pattern to see if the number goes up quickly. If so, it is probably multiplication.

If the numbers go down in the pattern, it is either subtraction or division.
Check the pattern to see if the number goes down quickly. If so, it is probably division.

Some cases involve addition and subtraction.
Some cases involve multiplication and division.

Some cases involve fractions.

Some cases involve positive and negative numbers.

You need to be prepared for all these different types of pattern and sequence problems.

Complete the following patterns.

Multiplication Pattern

1, 5, 25, 125, _____

Look for the pattern. When the numbers are getting larger, it is either addition or multiplication.
In this case, each number is multiplied by 5 to lead to the next number.
The next number would be 625.

Addition and Subtraction Pattern

4, 16, 8, 20, 12, _____

The pattern involves addition and subtraction.
The pattern is: add 12, subtract 8, add 12, subtract 8, add 12.
The next number would be 24.

Patterns with Fractions

1/2, 1/6, 1/18, _____

Each fraction is divided in 3 to lead to the next one. The denominator in each one is tripled.
The next number would be 1/54.

Pattern with Positive and Negative Numbers

6, -2, 4, -4, 2, -6, _____

The pattern involves addition and subtraction.
The pattern is: subtract 8, add 6, subtract 8, add 6.
The next number would be 0.

Pattern with Addition

9, 12, 17, 24, _____

The pattern is: add 3, add 5, add 7, add 9.
The next number would be 33.

Pattern with Subtraction

80, 70, 62, 56, _____

The pattern is: subtract 10, subtract 8, subtract 6, subtract 4.
The next number would be 52.

Arithmetic Reasoning
Patterns and Sequences - Pre-Test

Complete the following patterns.

1.

   1, 3, 9, 27, _____

2.

   5, 12, 10, 17, 15, _____

3.

   1/2, 1/4, 1/8, _____

4.

   5, -5, 0, -10, _____

5.

   1, 4, 9, 16, _____

6.

   100, 90, 82, 76, _____

Arithmetic Reasoning
Patterns and Sequences – Pre-Test with Answers

Complete the following patterns.

1.

1, 3, 9, 27, _____

Look for the pattern. In this case, each number is multiplied by 3 to lead to the next number. The next number would be <u>81</u>.

2.

5, 12, 10, 17, 15, _____

The pattern involves addition and subtraction. The pattern is: add 7, subtract 2, add 7, subtract 2, add 7. The next number would be <u>22</u>.

3.

1/2, 1/4, 1/8, _____

Each fraction is divided in 2 to lead to the next one. The denominator in each one is doubled. The next number would be <u>1/16</u>.

4.

5, -5, 0, -10, _____

The pattern involves addition and subtraction. The pattern is: subtract 10, add 5, subtract 10, add 5. The next number would be <u>-5</u>.

5.

1, 4, 9, 16, _____

The pattern is: add 3, add 5, add 7, add 9. The next number would be <u>25</u>.

6.

100, 90, 82, 76, _____

The pattern is: subtract 10, subtract 8, subtract 6, subtract 4. The next number would be
<u>72</u>.

Probability and Consecutive Integer Review

Probability

Probability is usually written as a fraction on the ASVAB.
The bottom number represents the total number of whatever is being selected from.
The top number represents what you are hoping to choose out of the total.

Example:
If there are 4 pink balloons and 3 blue balloons, what is the probability of picking a pink balloon out of all the balloons if one balloon is chosen?
pink balloons/total balloons = 4/7

Common types of probability questions could involve a deck of cards or rolling dice.

Probability in Rolling Dice

- **Be sure to know that a die has 6 sides: 1, 2, 3, 4, 5, 6.**

Example:

If Charlotte rolls a die, what is the probability on a number less than 4?

There are 6 sides on a die and 3 of them are less than 4.
The probability of rolling a die and landing on a number less than 4 is
3/6 = 1/2

Deck of Cards

- **Be sure to know there are 52 cards in a deck.**

A deck of cards could be used in a probability question.
Be sure to know how many cards are in a deck and the details of the cards.
There are 52 cards in a deck.
There are 4 suits: hearts, clubs, diamonds, and spades.
There are 13 cards in each suit: A, 2, 3, 4, 5, 6, 7, 8, 9, 10, Jack, Queen, King.
Each of the 13 cards comes in 4 suits.
For example, there is an ace of hearts, an ace of clubs, an ace of diamonds, and an ace of spades.

Probability in Choosing from a Deck of Cards

Example:
What is the probability of picking a spade out of a deck of cards?
To be able to solve probability questions about a deck of cards, you need to know how many cards are in a deck.

There is a total of 52 cards in a deck.
There are 4 suits: hearts, diamonds, clubs, and spades.
There are 13 cards for each suit: A, 2, 3, 4, 5, 6, 7, 8, 9, 10, Jack, Queen, and King.
There are 52 cards in a deck, and there are 13 spades in a deck.
The probability of picking a spade out of a deck of cards is:
spades/total = 13/52 = 1/4

Probability in Picking from Balloons

If there are 10 green balloons, 6 orange balloons, 4 purple balloons, and 5 black balloons in a bag, what is the probability of picking a green or an orange balloon from the bunch?

There are a total of 25 balloons, and there are 10 green and 6 orange balloons.

green + orange balloons/ total balloons = 10 + 6 /25 = 16/25

Consecutive Integers

There are questions that ask what three numbers in a row add up to a certain sum.
To find the answer, divide by 3 to find the middle of the three numbers.
Write down the number that comes before that number and the number that comes after that number.
You will have the three consecutive integers that add up to the given sum.

Here are examples of consecutive number questions.

Consecutive Integers
Example:
If there are three consecutive numbers with a sum of 336, what is the largest of the three numbers?

336 ÷ 3 = 112 is the middle number.
The three consecutive numbers are 111, 112, and 113.
The largest of the three numbers is 113.

Consecutive Even Integers

When dealing with finding three consecutive even integers, the three integers are even. Divide the sum of the three integers by 3. Find the middle number. Write down the even number before that number and the even number after that number. Skip the odd numbers. You will have found the three consecutive even integers that give you the original sum.

Example:
If there are three consecutive even numbers with a sum of 294, what is the second number of the three numbers?

$294 \div 3 = 98$ is the middle number.
The three numbers are consecutive even numbers, so we need to skip over the odd numbers.
The three consecutive even numbers are 96, 98, and 100.
The second of the three numbers is 98.

Consecutive Odd Numbers

When dealing with finding three consecutive odd integers, the three integers are odd. Divide the sum of the three integers by 3. Find the middle number. Write down the odd number before that number and the odd number after that number. Skip the even numbers. You will have found the three consecutive odd integers that give you the original sum.

Example:
If there are three consecutive odd numbers with a sum of 219, what is the smallest of the three numbers?

$219 \div 3 = 73$ is the middle number.
The three numbers are consecutive odd numbers, so we need to skip over the even numbers.
The three consecutive odd numbers are 71, 73, 75.
The smallest of the three numbers is 71.

Arithmetic Reasoning
Probability and Consecutive Integers - Pre-Test

1. If Oscar rolls a die, what is the probability of landing on a number less than 3?

2. What is the probability of picking a Queen out of a deck of cards?

3. If there are 8 pink balloons, 5 red balloons, 6 blue balloons, and 3 yellow balloons in a bag, what is the probability of picking a pink or a blue balloon from the bunch?

4. If there are 3 consecutive numbers with a sum of 156, what is the largest of the three numbers?

5. If there are 3 consecutive even numbers with a sum of 114, what is the second number of the three numbers?

6. If there are 3 consecutive odd numbers with a sum of 69, what is the smallest of the three numbers?

Arithmetic Reasoning
Probability and Consecutive Integers - Pre-Test with Answers

1. If Oscar rolls a die, what is the probability of landing on a number less than 3?

    There are 6 sides on a die and 2 of them are less than 3.
    The probability of rolling a die and landing on a number less than 3 is
    $2/6 = 1/3$

2. What is the probability of picking a Queen out of a deck of cards?

    To be able to solve probability questions about a deck of cards, you need to
    know how many cards are in a deck.

    There are a total of 52 cards in a deck.
    There are 4 suits: hearts, diamonds, clubs, and spades.
    There are 13 cards for each suit: A, 1, 2, 3, 4, 5, 6, 7, 8, 9, 10, Jack, Queen, and King.
    There are 52 cards in a deck and there are 4 Queens in a deck.
    The probability of picking a Queen out of a deck of cards is:

    Queens/total = $4/52 = 1/13$

3. If there are 8 pink balloons, 5 red balloons, 6 blue balloons, and 3 yellow balloons in a
    bag, what is the probability of picking a pink or a blue balloon from the bunch?

    There are a total of 22 balloons, and there are 8 pink and 6 blue balloons.
    pink + blue balloons/ total balloons = $8 + 6 / 22 = 14/22 = 7/11$

4. If there are 3 consecutive numbers with a sum of 156, what is the largest of
    the three numbers?

    $156 \div 3 = 52$ is the middle number.
    The three consecutive numbers are 51, 52, and 53.
    The largest of the 3 numbers is 53.

5. If there are 3 consecutive even numbers with a sum of 114, what is the second number of the three numbers?

    $114 \div 3 = 38$ is the middle number.
    The three numbers are consecutive even numbers, so we need to skip over the odd numbers.
    The three consecutive even numbers are 36, 38, and 40.
    The second of the three numbers is 38.

6. If there are 3 consecutive odd numbers with a sum of 69, what is the smallest of the three numbers?

    $69 \div 3 = 23$ is the middle number.
    The three numbers are consecutive odd numbers, so we need to skip over the even numbers.
    The three consecutive odd numbers are 21, 23, 25.
    The smallest of the three numbers is 21.

Distance and Rate Review

There are different types of distance and rate questions that can appear on the ASVAB. Many people are intimidated by the questions about trains traveling opposite directions, but when you see how to solve them, you will realize these questions are much simpler than they seem. You will be relieved!

Trains Traveling Opposite Directions

When a question asks about trains, cars, buses or planes traveling opposite directions, add up the distances that each one travels in one hour. This will show how far apart they are after one hour. Then multiply by the number of hours they travel.

Example:
Two trains are in the same city and start traveling in opposite directions. If one train travels east at 50 mph and the other train travels west at 20 mph, how far apart will the trains be at the end of 3 hours?

When trains travel opposite directions, add the distances each travel in an hour and multiply by the number of hours.
50 + 20 = 70
70 x 3 = 210 miles

Trains Traveling Toward Each Other

When a question asks about trains, cars, buses or planes traveling toward each other, they are still traveling opposite directions from each other. Add the distances traveled by each one in an hour. If presented with the total distance traveled, divide the distance traveled by both in one hour into the total distance to see how many hours it takes to travel the total distance.

Example:
If there are two planes that are 2400 miles apart and they are traveling toward each other, how long will it take the two planes to meet if one is traveling at 340 mph and the other is traveling at 260 mph?

When the planes are traveling toward each other, add the distances traveled in one hour by each plane and divide into the total distance to be traveled.
340 + 260 = 600
2400 ÷ 600 = 4 hours

Trains Traveling the Same Direction

When trains or any other vehicles are traveling the same direction, subtract their speeds to see how far ahead the faster train is in one hour. Multiply that distance by the total number of hours traveled. Then you know the total distance ahead that one train is over another.

Example:
If two trains are traveling in the same direction, and one is traveling at a speed of 70 mph, and the other is traveling at a speed of 40 mph, how far apart will the trains be at the end of 4 hours?

When cars travel the <u>same</u> direction, <u>subtract</u> the two speeds to see how far ahead one is after an hour and then <u>multiply</u> by the total number of hours.
70 – 40 = 30 mph
30 x 4 = 120 miles

Time of Arrival at Destination

When given a total distance traveled and the speed per hour, divide the speed into the total distance to figure out how many hours the trip took. To figure out the time of arrival at the destination, count on that many hours from the starting time.

Example:
A train travels a total distance of 490 miles. It travels at a speed of 70 mph. If the train leaves at 10 AM, when will the train reach its destination?

490 ÷ 70 = 7 hours
7 hours after 10 AM is 5 PM

Finding Average Speed of Car Traveling at Varying Rates of Speed

When a problem asks about average speed of a car that travels at various rates of speed, figure out the total distance traveled by the car. Then divide by the total number of hours traveled.

Example:
If a car travels at a rate of 60 mph for 5 hours and travels at a rate of 40 mph for 3 hours, what was the average rate of speed in miles per hour?

(60 x 5) + (40 x 3) =
300 + 120 = 420 miles
The total hours traveled was 5 + 3 = 8 hours
420 ÷ 8 = 52.5 mph

In a question like this, the car travels at 60 miles per hour for a period of time and 40 mph for a different period of time. The trap answer would be to add up 40 + 60 and divide by 2 to get 50 mph. A question like this is not that simple. Do not fall for this trap!

Man Forgets His Cell Phone

This question is a challenge for a number of people.
When a question asks about someone forgetting a cell phone or other item, calculate the total distance traveled by that person. The total distance tells you how far the person must travel to catch up with the original person.

A young man leaves home to head for college and forgets his cell phone. He drives at a speed of 50 mph. His father realizes he forgot his cell phone and follows him in his car 1 hour after he left. If he wants to catch up with him 3 hours after he left, how fast does the father have to drive?

The son leaves and travels 50 mph in the first hour. The father wants to catch up with his 3 hours after the son left.
50 x 3 = 150
The son will have traveled 150 miles.
His father needs to cover 150 miles in 2 hours.
150 ÷ 2 = 75 mph
The father will need to travel at 75 mph.

Steps:
1. Multiply hours by the son's speed.
   50 x 3 = 150 miles
2. Subtract hours son will travel by how many hours later the father left.
   3 – 1 = 2 hours
3. Divide Step 1 by Step 2.
   150 ÷ 2 = 75 mph
   The father will need to travel at 75 mph.

Arithmetic Reasoning
Distance and Rate - Pre-Test

1. Two trains are in the same city and start traveling in opposite directions. If one train travels east at 70 mph and the other train travels west at 40 mph, how far apart will the trains be at the end of 5 hours?

2. If there are two planes that are 4,000 miles apart and they are flying toward each other, how long will it take the two planes to meet if one is traveling at 420 mph, and the other is traveling at 380 mph?

3. If two cars are traveling in the same direction, and one is traveling at a speed of 30 mph, and the other is traveling at a speed of 50 mph, how far apart will the cars be at the end of 5 hours?

4. A train travels a total distance of 540 miles. It travels at a speed of 60 mph. If the train leaves at 11 AM, when will the train reach its destination?

5. If a car travels at a rate of 55 mph for 4 hours and travels at a rate of 45 mph for 3 hours, what was the average rate of speed in miles per hour?

6. A young man leaves home to head for college and forgets his wallet. He drives at a speed of 60 mph. His mother realizes he forgot his wallet and follows him in her car 1 hour after he left. If she wants to catch up with him 3 hours after he left, how fast does she have to drive?

Arithmetic Reasoning
Distance and Rate - Pre-Test with Answers

1. Two trains are in the same city and start traveling in opposite directions. If one train travels east at 70 mph and the other train travels west at 40 mph, how far apart will the trains be at the end of 5 hours?

   When trains move opposite directions, add the distances each travel in an hour and multiply by the number of hours.
   70 + 40 = 110
   110 x 5 = 550 miles

2. If there are 2 planes that are 4000 miles apart and they are flying toward each other, how long will it take the two planes to meet if one is traveling at 420 mph and the other is traveling at 380 mph?

   Add the distances traveled in one hour by each plane and divide into the total distance to be traveled.
   380 + 420 = 800
   4000 ÷ 800 = 5 hours

3. If 2 cars are traveling in the same direction, and one is traveling at a speed of 30 mph, and the other is traveling at a speed of 50 mph, how far apart will the cars be at the end of 5 hours?

   When cars travel the same direction, subtract the 2 speeds to see how far ahead one is after an hour and then multiply by the total number of hours.
   50 – 30 = 20 mph
   20 x 5 = 100 miles

4. A train travels a total distance of 540 miles. It travels at a speed of 60 mph. If the train leaves at 11 AM, when will the train reach its destination?

   540 ÷ 60 = 9 hours
   9 hours after 11 AM is 8 PM.

5.   If a car travels at a rate of 55 mph for 4 hours and travels at a rate of 45 mph for 3 hours, what was the average rate of speed in miles per hour?

(55 x 4) + (45 x 3) =
220 + 135 = 355 miles
The total hours traveled was 4 + 3 = 7 hours
355 ÷ 7 = 50.71 mph

6.   A young man leaves home to head for college and forgets his wallet. He drives at a speed of 60 mph. His mother realizes he forgot his wallet and follows him in her car 1 hour after he left. If she wants to catch up with him 3 hours after he left, how fast does the mother have to drive?

The son leaves and travels 60 mph in the first hour. The mother wants to catch up with him 3 hours after he leaves.
The son will have traveled 180 miles.
She needs to cover 180 miles in 2 hours.
180 ÷ 2 = 90 mph

Steps:
1. Multiply hours by the son's speed.
   60 x 3 = 180 miles
2. Subtract hours son will travel by how many hours later the mother left.
   3 − 1 = 2 hours
3. Divide Step 1 by Step 2.
   180 ÷ 2 = 90 mph
   The mother will need to travel at 90 mph.

Miles Per Gallon and Algebra in Word Problems Review

Miles per gallon questions are <u>NOT</u> proportion problems. People often mistake them for proportion problems.

Each mile per gallon problem has 3 numbers in the problem. There are 2 situations that can happen. Either 2 of the values are money, and one is a regular number; or 2 of the values are regular numbers, and one is a money value.

In this example, there is one money value and 2 regular numbers.
Divide number into number and multiply by the money value to solve.

Example:

Ralph buys gas for $4.70 a gallon. His car gets 30 miles per gallon. How much will a trip of 1200 miles cost?

This problem has 2 steps in order to solve it. To help you remember the steps, remember that in this problem there are 3 amounts mentioned: $4.70, 30, and 1200. 1 of them is a money amount, and the other 2 are numbers.

Divide number into number and multiply by money.
$1200 \div 30 = 40$

Then multiply by $4.70
$4.70 \times 40 = \$188.00$

In this example, there are 2 money values and 1 regular number.
Divide money into money and multiply by the regular number to solve.

Example:

Elizabeth spent $92.00 on gas, and gas costs $4.60 per gallon. Her car gets 45 miles per gallon. How far did she drive?

This problem has 2 steps in order to solve it. To help you remember the steps, remember that in this problem there are 3 amounts mentioned: $92.00, $4.60, and 45. 2 of them are money amounts, and 1 is a number.

Divide money into money and then multiply by number.
$\$92.00 \div \$4.60 = 20$
$20 \times 45 = 900$ miles

Solving Word Problems with Algebra

Some word problems can be solved using algebra.
It is helpful at times to use the first initial of the names of the people as the variable in the problem.

Example:
Jesse and Kyle have the same number of toy cars. Jesse has 5 times as many toy cars Martin, and Kyle has 24 more cards than Martin. How many cards does Martin have?

Translate into algebra.
Jesse has the same number of toy cars as Kyle.
$J = K$

Jesse has 5 times as many toy cars as Martin.
$J = 5M$

Kyle has 24 more toy cars than Martin.
$K = M + 24$

Since $J = K$, it is also true that $5M = M + 24$

Use $5M = M + 24$ to solve for M.
$4M = 24$
$M = 6$

Martin has 6 toy cars.

To check the work, plug back into the original equation.

$J = K$
$J = 5M = 5(6) = 30$
Jesse has 30 toy cars.

$K = M + 24 = 6 + 24 = 30$
Kyle has 30 toy cars.
Therefore, it does work that $J = K$ when $M = 6$.

Example:
Daniel is 10 years older than Brandon, and Brandon is 2 years younger than Eric. If the sum of their ages is 57, how old is Brandon?

Translate into algebra.

Daniel = Brandon + 10
D = B + 10

Brandon is 2 years younger than Eric which means Eric is 2 years older than Brandon.
B = E - 2
E = B + 2
Brandon = ?
B = ?

Daniel's age + Eric's age + Brandon's age = 57
(B + 10)   +   (B + 2)   +   B   =   57

Add up the terms.
3B + 12 = 57
Solve for B.
3B = 45
B = 15
Brandon is 15 years old.

To check your work, plug 15 back into the equation.

(B + 10) + (B + 2) + B = 57
(15 + 10) + (15 + 2) + 15 = 57
25 + 17 + 15 = 57
57 = 57

Daniel is 25, Eric is 17, and Brandon is 15, and it checks to be correct.

Spending Fractional Amount of Total

Example:
A man spends 1/6 of his income on his rent and 1/4 of the remainder of his income on his food. What part of his salary does he spend on food?

After a man spends 1/6 of his income on his car payment, he has 5/6 of his income left. He spends 1/4 of his remaining money on his food.
The word "of" means multiplication, so multiply 1/4 by 5/6 to find the answer.
1/4 x 5/6 = 5/24
The man spends 5/24 of his income on food.

Distance Traveled in Taxicab Question

Example:
Franklin rode in a taxi and his trip cost $68.00. The initial mile cost $2.00, and the trip cost $1.50 each additional ½ mile. How many miles long was Franklin's ride in the taxi?

68.00 – 2.00 (for initial mile) = 66.00

The trip cost $1.50 for each additional 1/2 mile so it cost $3.00 per mile.

Division will show how many additional miles Franklin rode in the taxi.
$66.00 ÷ $3.00 = 22
Add the first mile to the 22 miles. The total trip was 22 + 1 = 23 miles.

Finding Missing Test Grade in Average Question

Example:
Betsy earns the following test grades: 82, 70, 65, and 91. What score will she need on the fifth test so that she ends up with an average of 80 in the class?

To end up with an average of 80 in the class from 5 tests, Betsy needs a total of 80 x 5 points on all the tests combined.
80 x 5 = 400

All the tests grades added up plus the missing test grade will total 400.

82 + 70 + 65 + 91 + x = 400

Then add up the 4 test grades that Betsy already received.

82 + 70 + 65 + 91 = 308

308 + x = 400

Subtract 308 from each side to figure out the test grade she must receive on the 5th test to end up with an 80 average.

$$
\begin{array}{rcl}
308 + x &=& 400 \\
-308 & & -308 \\
\hline
x &=& 92
\end{array}
$$

Arithmetic Reasoning
Miles Per Gallon, Algebra in Word Problems - Pre-Test

1. Carl buys gas for $4.50 a gallon. His car gets 20 miles per gallon. How much will a trip of 600 miles cost?

2. Xavier spent $96.00 on gas, and gas costs $4.80 per gallon. His car gets 33 miles per gallon. How far did he drive?

3. Derek and Kieran have the same number of baseball cards. Derek has 4 times as many baseball cards as Brian, and Kieran has 15 more cards than Brian. How many cards does Brian have?

4. Haley is 7 years older than Clara, and Clara is 1 year younger than Alyssa. If the sum of their ages is 35, how old is Clara?

5. A man spends 1/4 of his income on his car payment and 1/3 of the remainder of his income on food. What part of his salary does he spend on food?

6. Esther rode in a taxi, and her trip cost $32.00. The initial mile cost $2.00, and the trip cost $.75 each additional 1/2 mile. How many miles long was Esther's ride in the taxi?

7. Jeremy earns the following test grades: 85, 75, 82, and 100. What score will he need on the fifth test so that he ends up with an average of 87 in the class?

Arithmetic Reasoning
Miles per gallon, Algebra in Word Problems - Pre-Test with Answers

1. Carl buys gas for $4.50 a gallon. His car gets 20 miles per gallon. How much will a trip of 600 miles cost?

    This problem has 2 steps in order to solve it. To help you remember the steps, remember that in this problem there are 3 amounts mentioned: $4.50, 20, and 600. 1 of them is a money amount, and the other 2 are numbers.

    Divide number into number and multiply by money.

    $600 \div 20 = 30$ then multiply by $4.50

    4.50 x 30 = $135.00

2. Xavier spent $96.00 on gas, and gas costs $4.80 per gallon. His car gets 33 miles per gallon. How far did he drive?

    This problem has 2 steps in order to solve it. To help you remember the steps, remember that in this problem there are 3 amounts mentioned: $96.00, $4.80, and 33.
    2 of them are money amounts, and 1 is a number.

    Divide money into money and then multiply by number.

    $96.00 \div $4.80 = 20

    20 x 33 = 660 miles

3. Derek and Kieran have the same number of baseball cards. Derek has 4 times as many baseball cards as Brian, and Kieran has 15 more cards than Brian. How many cards does Brian have?

    Translate into algebra.

    Derek has the same number of cards as Kieran.
    D = K

102

Derek has 4 times as many baseball cards as Brian.
D = 4B

Kieran has 15 more cards than Brian.
K = B +15

D = K
D = 4B and K = B +15
4B = B + 15

Solve for B.
3B = 15
B = 5
Brian has 5 baseball cards.

To check your work, plug 5 back into the equation.
D = 4B = 4(5) = 20
K = B + 15 = 5 + 15 = 20

Derek has 20 baseball cards and Kieran has 20 baseball cards so it checks out to be true that Derek and Kieran have the same number of baseball cards.
Therefore, it does work that D = K when B = 5.

4.   Haley is 7 years older than Clara, and Clara is 1 year younger than Alyssa. If the sum of their ages is 35, how old is Clara?

Translate into algebra.

Haley = Clara + 7
H = C + 7

Clara is 1 year younger than Alyssa.
C = A - 1

This also means Alyssa is 1 year older than Clara.
A = C+1

This formula, A = C+1, is easier to solve for than C = A-1 since we need to figure out Clara's age.
C = ?

Haley's age + Alyssa's age + Clara's age = 35
(C + 7) + (C + 1) + C = 35

3C + 8 = 35
3C = 27
C = 9
Clara is 9 years old.

To check your work, plug 9 back into the equation.

H   +   A +   C = 35
(C + 7) + (C + 1) + C = 35
(9 + 7) + (9 + 1) + 9 = 35
16 + 10 + 9 = 35
35 = 35

Haley is 16, Alyssa is 10, and Clara is 9 and it checks to be correct.

5.   A man spends 1/4 of his income on his car payment and 1/3 of the remainder of his income on food. What part of his salary does he spend on food?

After a man spends 1/4 of his income on his car payment, he has 3/4 of his income left. He spends 1/3 of his remaining money on his food.
1/3 x 3/4 = 3/12 = 1/4

6.   Esther rode in a taxi and her trip cost $32.00. The initial mile cost $2.00, and the trip cost $.75 each additional 1/2 mile. How many miles long was Esther's ride in the taxi?

$32.00 – $2.00 (for initial mile) = $30.00

The trip cost $.75 for each additional 1/2 mile so it cost $1.50 per mile.

Division will show how many additional miles Esther rode in the taxi.
$30.00 ÷ $1.50 = 20

Add the first mile to the 20 miles. The total trip was 20 + 1 = 21 miles.

7.   Jeremy earns the following test grades: 85, 75, 82, and 100. What score will he need on the fifth test so that he ends up with an average of 87 in the class?

There are a total of 5 tests so 87 needs to be multiplied by 5 to figure out the total number of points needed to earn an average of 87.

$$85 + 75 + 82 + 100 + x = 87 (5)$$

$$85 + 75 + 82 + 100 + x = 435$$

$$342 + x = 435$$

$$
\begin{aligned}
342 + x &= 435 \\
-342 &\quad -342 \\
\hline
x &= 93
\end{aligned}
$$

He needs a score of 93.

<h1 style="text-align:center">Measurement Review</h1>

Units of Measurement

Distance or Length

1 foot = 12 inches

1 yard = 3 feet = 36 inches

1 mile = 1760 yards = 5280 feet

Weight

1 pound = 16 ounces

1 ton = 2000 pounds

Other Measurements

1 dozen = 12

1 gross = 144 = 12 dozen

1 score = 20

Measurement Questions Involving Distance and Length

If a 6 yd piece of wood is cut into 8 pieces, how long is each piece?

To solve this problem, you need to know that there are 3 ft in 1 yard, and there are 12 inches in 1 ft.

6 yd/8 = 18 ft/8 = 216 inches/8 = 27 inches = 2 ft 3 inches

Example:

If a 15 ft 8 inch piece of wood needs to be divided into 4 pieces, how long would each piece be?

To solve this problem, you need to know there are 12 inches in a foot.

15 ft 8 inches = (15 x 12) + 8 = 180 + 8 = 188 inches
188 ÷ 4 = 47 inches or 3 ft 11 inches

Example:

If Joe ran 7 miles and Ron ran 2 miles, how many more yards did Joe run than Ron?

To solve this problem, you need to know that there are 1760 yards in a mile.

7 – 2 = 5 miles

There are 1,760 yards in a mile. 1760 x 5 = 8,800 yards

Example:

If Zoe walked 1 mile in 30 minutes, how many feet did she walk per minute?

To solve this problem, you need to know there are 5,280 feet in a mile.

There are 5,280 feet in a mile.

5280 ÷ 30 = 176 feet in a minute

Example:

If a room measures 12 ft by 15 ft, and carpet cost $3.50 per square yard, how much would it cost to carpet the room?

To solve this problem, you need to know that there are 3 feet in 1 yard.

The carpet cost is measured in square yards.

12 ft x 15 ft change into yards by dividing each one by 3.

4 yd x 5 yd = 20 square yards = 20 yd^2

$3.50 x 20 = $70.00

Measurement Questions Involving Weight

Example:
Gianna's baby was born weighing 7 lb 10 oz and Lena's baby was born weighing 5 lb 11 oz.
How much more did Gianna's baby weigh than Lena's baby at birth?
To solve this problem, you need to know there are 16 ounces in a pound.

There are two ways to solve this problem.

a. Change the pounds to ounces and subtract.

7 lb 10 oz = (7 x 16) + 10 = 112 + 10 = 122 oz
5 lb 11 oz = (5 x 16) + 11 = 80 + 11 = 91 oz
122 – 91 = 31 oz
31 ÷ 16 = 1 lb with 15 oz left over
1 lb 15 oz

or

b. Borrow from the pounds to be able to subtract the ounces.

7 lb 10 oz – 5 lb 11 oz
Since you cannot take 11 oz away from 10 oz,
you need to turn the 7 lb into 6 lb and add 16 oz to the ounces.
Then subtract.
6 lb 26 oz – 5 lb 11 oz = 1 lb 15 oz

Example:

Emily orders 120 pounds of hamburger meat. How many 1/4 pound burgers can she make using 120 pounds of meat?

Emily has 120 pounds of hamburger meat that needs to be divided into 1/4 pound hamburgers.

120 ÷ 1/4 =

120/1 x 4/1 = 480 hamburgers

Example:

If a truck carries 4 tons of cat food and each cat receives a daily food quota of 4 ounces. How many cats can be fed from this delivery for a total of 10 days?

To solve this problem, you need to know that 1 ton = 2000 pounds and 1 pound = 16 ounces.

4 tons = 8000 pounds
Each cat eats 4 ounces a day, which is 40 ounces in 10 days.
40 ounces = 2.5 pounds
$8,000 \div 2.5 = 80,000 \div 25 = 3,200$ cats

Another way to solve it is:
4 tons = 8,000 pounds
Each cat eats 4 ounces a day, which is 40 ounces in 10 days.
Change 8000 pounds to ounces.
$8000 \times 16 = 128,000$ ounces
$128,000 \div 40 = 3,200$ cats

Time

60 seconds = 1 minute
60 minutes = 1 hour
24 hours = 1 day
7 days = 1 week
52 weeks = 1 year
365 days = 1 year
10 years = 1 decade
100 years = 1 century
1,000 years = 1 millennium

Months of the Year
Be sure to know how many days are in all the months of the year.

There is a poem to remember how many days are in each month.

Thirty days hath September, April, June, and November.
All the rest have thirty-one.
February has twenty-eight,
But leap year coming once in four.
February then has one day more.

January - 31 days
February - 28 days, except for Leap Year when it has 29 days once every 4 years.
March - 31 days
April - 30 days
May - 31 days
June - 30 days
July - 31 days
August - 31 days
September - 30 days
October - 31 days
November - 30 days
December - 31 days

Metric Measurement
centi - hundredth
100 cm = 1 meter

milli - thousandth
1000 mm = 1 meter
1000 mg = 1 gram

kilo - thousand
1 kilometer = 1000 meters
1 kilogram = 1000 grams

Metric Measurement and Time Question

Example:
Lorenzo ran 750 meters per day. How many km would he have run in the months of March, April, and May?

To solve this problem, you need to know there are 1000 meters in a kilometer.

You also need to know how many days are in each month of the year.

January = 31 days
February = 28 days, except on Leap Year when it has 29 days once every 4 years.
March = 31 days
April = 30 days
May = 31 days
June = 30 days
July = 31 days
August = 31 days
September = 30 days
October = 31 days
November = 30 days
December = 31 days

There is also an old poem to help remember the days in the month.

Thirty days hath September, April, June, and November.
All the rest have thirty-one.
February has twenty-eight,
But leap year, coming once in four.
February then has one day more.

March has 31 days, April has 30, and May has 31 days.

31 + 30 + 31 = 92 days
92 x 750 = 69,000 meters

There are 1000 meters in a kilometer.

69,000 ÷ 1000 = 69 km

Make sure you remember how many days are in each of the months of the year.

Metric Measurement Questions

Example:

If a sunflower plant grows 8 mm per day, how many meters will it grow in 1 year?

To solve this problem, you need to know that there are 1000 mm in a meter.
You also need to know that there are 365 days in a year.

8 mm x 365 = 2920 mm
There are 1000 mm in a meter.
2920 ÷ 1000 = 2.92 meters

Example:
Rosa ran 2400 meters and Christine ran 5 km. How much farther did Christine run than Rosa?

To solve this problem, you need to know there are 1000 meters in a kilometer.

Rosa ran 2400 meters, and Christine ran 5 km, which equals 5000 meters.
5000 − 2400 = 2600 meters
If the answer needed to be written in kilometers, divide by 1000.
2600 meters = 2.6 km

Liquid Measurement
Gallon Man
This illustration will help you to remember liquid measurements of :
cup (C), pint (P), quart (Q), and gallon (G).

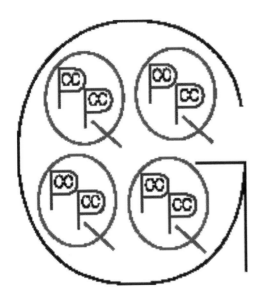

1 gallon = 4 quarts
1 quart = 2 pints
1 pint = 2 cups
1 cup = 8 ounces

Liquid Measurement Questions

Example:
A cup of milk costs $.50, and a gallon size container costs $4.00. How much would someone save by buying 2 gallons in gallon-sized containers instead of buying the same amount of milk in cup-sized containers?

To solve this problem, you need to know there are 16 cups in a gallon.

There are 16 cups in a gallon, and there are 32 cups in 2 gallons.
$.50 x 32 = $16.00

2 gallons would cost $4.00 x 2 = $8.00

$16.00 - $8.00 = $8.00 more by buying cup-sized containers instead of the gallon-sized containers.

A person would save $8.00 by buying the gallon-sized containers.

Example:

A swimming pool has dimensions of 30 ft long by 15 ft wide by 5 ft deep. A cubic foot of water is about 7.5 gallons. How many gallons would it take to fill the pool?

To solve this problem, you need to know the volume of a rectangular box.
V = L x W x H

Find the volume of the pool.
L x W x H
30 x 15 x 5 = 2,250 cubic ft. = 2,250 ft^3

Multiply the volume by the number of gallons per cubic foot.
2250 x 7.5 = 16,875 gallons

Arithmetic Reasoning
Measurement - Pre-Test

1.   If an 8 yd piece of wood is cut into 10 pieces, how long is each piece?

2.   If 13 ft 8 inches needs to be divided into 4 pieces, how long would each piece be?

3.   If Tom ran 5 miles and Richard ran 3 miles, how many more yards did Tom run than Richard?

4.   If Lizzy walked 1 mile in 20 minutes, how many feet did she walk per minute?

5.   If a room measures 9 ft by 12 ft, and carpet cost $5.50 per square yard, how much would it cost to carpet the room?

6.   A pint of milk costs $.75, and a gallon-sized container costs $5.00. How much would someone save by buying 2 gallons in gallon-sized containers instead of buying the same amount of milk in pint-sized containers?

7   A swimming pool has dimensions of 40 ft long by 10 ft wide by 4 ft deep. A cubic foot of water is about 7.5 gallons. How many gallons are used to fill the pool?

8.   Jean's baby was born weighing 8 lb 1 oz, and Connie's baby was born weighing 6 lb 7 oz. How much more did Jean's baby weigh than Connie's baby at birth?

9.   Christopher orders 80 pounds of hamburger meat. How many 1/4 pound burgers can he make using 80 pounds of meat?

10.   If a truck carries 3 tons of cat food and each cat receives a daily food quota of 8 ounces. How many cats can be fed from this delivery for a total of 10 days?

11.   June ran 1900 meters and Marissa ran 2 km. How many meters farther did Marissa run than June?

12.   If a sunflower plant grows 10 mm per day, how many meters will it grow in 1 year?

13.   If Jackie ran 500 meters per day, how many km would she have run in the months of September, October, and November?

Julie A. Hyers ©2020

Arithmetic Reasoning
Measurement - Pre-Test with Answers

1. If an 8 yd piece of wood is cut into 10 pieces, how long is each piece?

    8 yd/10 = 24 ft/10 = 288 inches/10 = 28.8 inches = 2 ft 4.8 inches

2. If 13 ft 8 inches needs to be divided into 4 pieces, how long would each piece be?

    13ft 8 inches = 164 inches
    $164 \div 4 = 41$ inches or 3 ft 5 inches

3. If Tom ran 5 miles and Richard ran 3 miles, how many more yards did Tom run than Richard?

    $5 - 3 = 2$ miles

    There are 1760 yards in a mile. 1760 x 2 = 3520 yards

4. If Lizzy walked 1 mile in 20 minutes, how many feet did she walk per minute?

    There are 5280 feet in a mile.
    $5280 \div 20 = 264$ feet in a minute

5. If a room measures 9 ft by 12 ft, and carpet cost $5.50 per square yard, how much would it cost to carpet the room?

    The carpet cost is measured in square yards.
    9 ft x 12 ft change into yards by dividing each one by 3.
    3 yd x 4 yd = 12 square yards = 12 yd^2
    $5.50 x 12 = $66.00

6.   A pint of milk costs $.75, and a gallon-sized container costs $5.00. How much would someone save by buying 2 gallons in gallon-sized containers instead of buying the same amount of milk in pint-sized containers?

There are 8 pints in a gallon, and there are 16 pints in 2 gallons.
$.75 x 16 = $12.00
2 gallons would cost $5.00 x 2 = $10.00
$12.00 - $10.00 = $2.00 more to buy pint-sized containers instead of the gallon-sized containers.

7.   A swimming pool has dimensions of 40 ft long by 10 ft wide by 4 ft deep. A cubic foot of water is about 7.5 gallons. How many gallons are used to fill the pool?

Find the volume of the pool.
L x W x H
40 x 10 x 4 = 1600 cubic ft. = 1,600 ft^3
Multiply the volume by the number of gallons per cubic foot.
1600 x 7.5 = 12,000 gallons

8.   Jean's baby was born weighing 8 lb 1 oz and Connie's baby was born weighing 6 lb 7 oz. How much more did Jean's baby weigh than Connie's baby at birth?

There are two ways to solve this problem.
  a. Change the pounds to ounces and subtract.

8 lb 1 oz – 6 lb 7 oz
8 lb 1 oz = (8 x 16) + 1 = 128 + 1 = 129 oz
6 lb 7 oz = (6 x 16) + 7 = 96 + 7 = 103 oz
129 – 103 = 26 oz
26 ÷ 16 = 1 lb 10 oz

    or

  b. Borrow from the pounds to be able to subtract the ounces.

8 lb 1 oz – 6 lb 7 oz
Since you cannot take 7 oz away from 1 oz, you need to turn 8 lb into 7 lb and add 16 oz to the ounces.
7 lb 17 oz – 6 lb 7 oz = 1 lb 10 oz

9. Christopher orders 80 pounds of hamburger meat. How many 1/4 pound burgers can he make using 80 pounds of meat?

$80 \div 1/4 =$
$80/1 \times 4/1 = 320$ hamburgers

10. If a truck carries 3 tons of cat food and each cat receives a daily food quota of 8 ounces, how many cats can be fed from this delivery for a total of 10 days?

3 tons = 6000 pounds
Each cat eats 8 ounces a day, which is 80 ounces in 10 days.
80 ounces = 5 pounds
$6000 \div 5 = 1,200$ cats
or
6000 pounds = 96,000 ounces
8 ounces a day = 80 ounces in 10 days
$96,000 \div 80 = 1,200$ cats

11. June ran 1900 meters and Marissa ran 2 km. How much farther did Marissa run than June?

June ran 1900 meters, and Marissa ran 2 km, which equals 2000 meters.
$2000 - 1900 = 100$ meters

12. If a sunflower plant grows 10 mm per day, how many meters will it grow in 1 year?

There are 365 days in a year.
$10mm \times 365 = 3650$ mm
There are 1000 mm in a meter
$3650 \div 1000 = 3.65$ meters

13. If Jackie ran 500 meters per day, how many km would she have run in the months of September, October, and November?

September has 30 days, October has 31, and November has 30 days.
$30 + 31 + 30 = 91$ days
$91 \times 500 = 45,500$ meters
$45500 \div 1000 = 45.5$ km

Post-Tests

Arithmetic Reasoning
Basic Operations - Post-Test

1. 9,347
    + 3,865

2. 40,000
    - 26,432

3. 346
    x 7

4. 748
    x 86

5. 534
    x 389

6. $973 \div 6 =$

7. $7,246 \div 48 =$

8. $93,456 \div 314 =$

Arithmetic Reasoning
Basic Operations - Post-Test with Answers

1.   9,347
    + 3,865
       13,212

2.   40,000
    - 26,432
       13,568

3.   346
    x 7
      2,422

4.   748
    x   86
      4488
    + 59840
      64,328

5.   534
    x 389
      4806
      42720
    + 160200
     207,726

6. $973 \div 6 =$

162 remainder 1

$$6 \overline{)\ 973}$$

$\underline{-6}$
37
$\underline{-36}$
     13
     $\underline{-12}$
          1

7. $7{,}246 \div 48 =$

150 remainder 46

$$48 \overline{)\ 7{,}246}$$

$\underline{-\ 48}$
     244
     $\underline{-\ 240}$
          46

8. $93{,}456 \div 314 =$

297   remainder 198

$$314 \overline{)\ 93{,}456}$$

$\underline{-\ 628}$
     3065
     $\underline{-\ 2826}$
          2396
          $\underline{-2198}$
               198

Arithmetic Reasoning
Decimals - Post-Test

1. Add decimals.
    $4.13 + 0.0032 + 300 =$

2. Subtract decimals.
    $734.9 - 0.006 =$

3. Multiply decimals.
    $500.35 \times .15 =$

4. Divide decimals.
    $250 \div .50 =$

Julie A. Hyers ©2020

Arithmetic Reasoning
Decimals - Post-Test with Answers

1. Add decimals.
    $4.13 + 0.0032 + 300 =$ Line up the decimal places before adding.

    ```
      4.1300
      0.0032
    + 300.0000
      304.1332
    ```

2. Subtract decimals.
    $734.9 - 0.006 =$ Line up the decimal places before subtracting.

    ```
      734.900
    -   0.006
      734.894
    ```

3. Multiply decimals.
    $500.35 \times .15 =$
    In this case, there are two digits after the decimal place in 500.35 and two digits after the decimal place in .15. $2 + 2 = 4$, and the decimal place in the answer needs to be moved four places to the left in the answer 750525 to become 75.0525

    ```
        500.35
    x      .15
        250175
    +   500350
        75.0525
    ```

4. Divide decimals.
    $250 \div .50 = 25000 \div 50 =$
    When dividing by a decimal, move the decimal over to the right so it is behind the number and then move the decimal the same number of places in the other number.
    Then $250 \div .50 = 25000 \div 50 = 500$

    ```
                      500
    0.50 ) 250  =  50 ) 25000
                     -250
                       00
                     -  0
                       00
                     -  0
                        0
    ```

Arithmetic Reasoning
Fractions - Post-Test

1.   Change improper fraction to mixed number.

       8/5 =

2.   Change mixed number to improper fraction.

       5   2/4 =

3.   5/8 + 2/4 =

4.   6/7 – 1/5 =

5.   4/9 x 2/5 =

6.   7/11 ÷ 3/4 =

Arithmetic Reasoning
Fractions - Post-Test with Answers

1. Change improper fraction to mixed number.

   8/5 = 1   3/5

2. Change mixed number to improper fraction.

   5   2/4 = 22/4

3. 5/8 + 2/4 =

   Find a common denominator.
   Add the numerators. Reduce if necessary.
   5/8 + 4/8 = 9/8 = 1 1/8

4. 6/7 – 1/5 =

   Find a common denominator.
   Subtract the numerators and reduce if necessary.

   30/35 - 7/35 = 23/35

5. 4/9 x 2/5 = 8/45

   Multiply the numerators.
   Multiply the denominators.
   Reduce if necessary.

6. 7/11 ÷ 3/4 =

   Keep, change, flip.
   Multiply the numerators.
   Multiply the denominators.
   Reduce if necessary.

   7/11 x 4/3 = 28/33

127

Arithmetic Reasoning
Mixed Numbers and Comparing Fractions - Post-Test

1.   4 1/3
    + 2 1/5

2.   3 1/3
    - 1 4/6

3.   4 2/5 x 3 1/2 =

4.   7 1/3 ÷ 1 1/4 =

5.   Put the following fractions in order from least to greatest.
      1/3, 5/6, 2/5, and 3/8

Arithmetic Reasoning
Mixed Numbers and Comparing Fractions - Post-Test with Answers

1.   4 1/3
       + 2 1/5

Add the whole numbers and add the fractions.
Find a common denominator.

       4 5/15
    + 2 3/15
       6 8/15

2.   3 1/3
    -   1 4/6

Find a common denominator.

       3 2/6
    -   1 4/6

4/6 cannot be subtracted from 2/6
Borrow from the 3. Make the 3 into 2.
The 1 that is borrowed becomes 6/6.
Add 6/6 to 2/6 to become 8/6.
Then subtract.

       2 8/6
    -   1 4/6
       1 4/6 = 1 2/3

3.   4 2/5 x 3 1/2 =

   Change mixed numbers to improper fractions.
   Multiply the numerators.
   Multiply the denominators.
   Change improper fraction back to mixed number.
   Reduce if needed.
   4 2/5 x 3   1/2 =

   22/5 x 7/2 = 154/10 = 15 4/10 = 15 2/5

4.   7 1/3 ÷ 1   1/4 =

   Change mixed numbers to improper fractions.
   Keep first fraction.
   Change division to multiplication.
   Flip second fraction over.
   Multiply the numerators.
   Multiply the denominators.
   Change improper fraction back to mixed number.
   Reduce if needed.

   22/3 ÷ 5/4 =

   22/3 x 4/5 = 88/15 = 5 13/15

5.   Put the following fractions in order from least to greatest.
   1/3, 5/6, 2/5, and 3/8

   Change to decimals to compare.
   1/3 = .3333
   5/6 = .8333
   2/5 = .4000
   3/8 = .3750

   In order from least to greatest:
   1/3, 3/8, 2/5, 5/6

Julie A. Hyers ©2020

Arithmetic Reasoning
Percentage - Post-Test

1.   What percent of 60 is 40?

2.   What is 40% of 80?

3.   21 is 30% of what number?

4.   A plant sells for $60.00 and is on sale for $40.00, what is the percent discount?

5.   If a bracelet costs $800.00 and the price increases to $1200.00, what is the percent increase?

131

6.   If a bike rack for a car costs $150 and is on sale for 20% off and then an additional 10% is taken off, what is the final sale price?

7.   If 60 ounces of a drink consists of 25% juice and 20 ounces of water are added to the drink, what is the percentage of juice in the new mixture?

8.   The sale price of a cell phone is $300.00, which is 25% off the original price. What is the original price?

9.   In a group of 80 people, 30 are male, what percent are male?

10.   If a town's population increased by 40% from 1970 to 1980 and decreased by 25% from 1980 to 1990, what was the percent increase in population from 1970 to 1990?

Arithmetic Reasoning
Percentage - Post-Test with Answers

1. What percent of 60 is 40?

    is/of = %/100

    40/60 = x/100
    4000 = 60x
    400 = 6x
    66.66% = x

2. What is 40% of 80?

    is/of = %/100

    x/80 = 40/100
    100x = 3200
    x = 32

3. 21 is 30% of what number?

    is/of = %/100

    21/x = 30/100
    2100 = 30x
    210 = 3x
    70 = x

4. A plant sells for $60.00 and is on sale for $40.00. What is the percent discount?

    To solve this problem, you need to know the percent increase and decrease formula.

    difference/original = %/100

    60-40/60 = x /100
    20/60 = x/100
    2/6 = x/100
    1/3 = x/100

Cross-multiply and solve for x.
100 = 3x
33.33% discount = x

5. If a bracelet costs $800.00 and the price increases to $1200.00, what is the percent increase?

To solve this problem, you need to know the percent increase and decrease formula.

difference/original = %/100

1200 - 800/800 = x/100
400/800 = x/100
1/2 = x/100
Cross-multiply and solve for x.
2x = 100
x = 50% increase

6. If a bike rack for a car costs $150 and is on sale for 20% off and then an additional 10% is taken off, what is the final sale price?

The percentage reductions must be done one at a time.
Change percent to a decimal.
20% = .20
Multiply by .20 and subtract the result.
150 x .20 = 30
150 – 30 = 120

Change percent to a decimal.
10% = .10
Multiply by .10, and subtract the result.
120 x .10 = 12
120 – 12 = $108

7. If 60 ounces of a drink consists of 25% juice and 20 ounces of water are added to the drink, what is the percentage of juice in the new mixture?

Find how many ounces of juice are in the original amount.
60 x .25 = 15 ounces
Total liquid = 60 + 20 = 80

juice/total = %/100
15/80 = x/100
1500 = 80x
Reduce to 150 = 8x
18.75% = x

8. The sale price of a cell phone is $300.00, which is 25% off the original price. What is the original price?

$300 is 25% off the original price.
The original price is 100%.

When you subtract 25% from 100%, you are left with 75%.

That means $300 is 75% of the original price.

300/x = 75/100

30000 = 75x
$400 = x

You can also reduce.
300/x = 3/4
1200 = 3x
$400 = x

9. In a group of 80 people, 30 are male. What percent are male?

male/total = %/100

30/80 = %/100
30/80 = x/100
Reduce 30/80 to become 3/8
3/8 = x/100
300 = 8x
37.5% = x

135

10. If a town's population increased by 40% from 1970 to 1980 and decreased by 25% from 1980 to 1990, what was the percent increase in population from 1970 to 1990?

Assume the starting population is 100.
If the population increases 40% from 1970 to 1980, then it is 140 in 1980.
If the population decreased by 25% from 1980 to 1990, find 25% of 140
140 x .25 = 35
140 – 35 = 105

The original population was 100 and, in the end, it is 105, which is a 5% increase.

Arithmetic Reasoning
Percent, Decimal, Fraction Conversions - Post-Test

1.   Change fraction to decimal.

   $4/9 =$

2.   Change mixed number to decimal.

   $2 \quad 3/7 =$

3.   Change fraction to percent.

   $4/5 =$

4.   Change mixed number to percent.

   $5 \quad 1/4 =$

5.   Change decimal to fraction.

   $.58 =$

6.   Change decimal to fraction.

      3.38 =

7.   Change decimal to percent.

      .47 =

8.   Change decimal to percent.

      5.23 =

9.   Change percent to decimal.

      45% =

10.   Change percent to decimal.

      528% =

11.   Change percent to fraction.

78% =

12.   Change percent to fraction.

875% =

Arithmetic Reasoning
Percent, Decimal, Fraction Conversions - Post-Test with Answers

1. Change fraction to decimal.

    $4/9 = .44$

    Divide the numerator into the denominator. $4 \div 9 = .44$

2. Change mixed number to decimal.

    $2\ 3/7 = 2.428$

    The whole number 2 remains a whole number.
    Divide the denominator into the numerator.
    $3 \div 7 = .428$
    The answer is 2.428.

3. Change fraction to percent.

    $4/5 = 80\%$

    $4/5 = x/100$
    Cross-multiply and solve for x.
    $400 = 5x$
    $80\% = x$

4. Change mixed number to percent.

    $5\ 1/4 = 525\%$

    Change to an improper fraction and set up a proportion.
    $21/4 = x/100$
    Cross-multiply and solve for x.
    $2100 = 4x$
    $525\% = x$

5.   Change decimal to fraction.

.58 = 58/100 = 29/50

See how many digits are after the decimal point. In this case there are 2 digits after the decimal point, so put 58 over 100 since 100 has 2 zeros. Reduce if necessary.

6.   Change decimal to fraction.

4.38 = 4 38/100 = 4 19/50

Keep the 4 as a whole number. Check to see how many digits are after the decimal point. Since there are 2 digits after the decimal point, write 38 over 100, since 100 has 2 zeros. Reduce if necessary.

7.   Change decimal to percent.

.47 = 47%

Move the decimal point 2 places to the right.

8.   Change decimal to percent.

5.23 = 523%

Move the decimal point 2 places to the right.

9.   Change percent to decimal.

45% = .45

Move the decimal point 2 places to the left.

10.   Change percent to decimal.

528% = 5.28

Move the decimal point 2 places to the left.

11.   Change percent to fraction.

78% = 78/100 = 39/50

Place the number over 100, and reduce if necessary.

12.   Change percent to fraction.

875% = 875/100 = 8 75/100 = 8   3/4

Place the number over 100, and reduce if necessary.

Arithmetic Reasoning
Proportions - Post-Test

1.   If a tree is 20 ft tall and its shadow is 8 ft tall and a flagpole is 9 ft tall, how tall is the flagpole's shadow?

2.   On a map, if 1 inch represents 25 miles, how many miles does 3 1/4 inches represent?

3.   If an apple pie for 12 people has 4 1/2 cups of apples, how many cups of apples are needed for an apple pie for 16 people?

4.   If 2 lb 8 oz of sugar costs $4.00, how much would 4 oz of sugar cost?

5.   If 5 oz of coffee costs $1.50, how much would 5 lb of coffee cost?

Julie A. Hyers ©2020

Arithmetic Reasoning
Proportions Post - Test with Answers

1. If a tree is 20 ft tall and its shadow is 8 ft tall and a flagpole is 9 ft tall, how tall is the flagpole's shadow?

    object/shadow = object/shadow
    20/8 = 9/x
    Cross-multiply and solve for x.
    20x = 72
    x = 72/20 = 3 12/20 = 3 6/10 = 3 3/5 ft or 3.6 ft

2. On a map, if 1 inch represents 25 miles, how many miles does 3 1/4 inches represent?

    To make it easier, change 3 1/4 into a decimal 3.25
    inches/miles = inches/miles
    1/25 = 3.25/x
    x = 25 x 3.25 = 81.25 = 81 1/4 miles

3. If an apple pie for 12 people has 4 1/2 cups of apples, how many cups of apples are needed for an apple pie for 16 people?

    people/cups = people/cups
    12/4.5 = 16/x
    You can reduce the fractions.
    12 and 16 can both be divided by 4.
    3/4.5 = 4/x
    3x = 18
    x = 6 cups

4. If 2 lb 8 oz of sugar costs $4.00, how much would 4 oz of sugar cost?

    To solve this problem, you need to know that there are 16 ounces in 1 pound.
    2 lb 8 oz = 2 x 16 = 32 + 8 = 40 oz
    oz/cost = oz/cost
    40/4 = 4/x
    Reduce 40/4 to 10/1
    10/1 = 4/x

Cross-multiply and solve for x.
10x = 4
x = $0.40

5. If 5 oz of coffee costs $1.50, how much would 5 lb of coffee cost?

To solve this problem, you need to know there are 16 ounces in 1 pound.
ounces/cost = ounces/cost
Change 5 lb to ounces. There are 16 ounces in a pound, so 5 lb is 80 ounces.
5/1.50 = 80/x
Reduce 5 and 80 to become 1 and 16
1/1.50 = 16/x
Cross-multiply and solve for x.
x = $24.00

Arithmetic Reasoning
Ratios - Post-Test

1. If there is a 9:4 ratio of cows to horses, and there are 63 cows, how many horses are there?

2. If a woman donates $100,000 to charity and she gives her money to a school and a community center in a 3:1 ratio, how much money did she give to the school?

3. If there are 20 boys and 15 girls playing in the park, what is the ratio of boys to children in the park?

4. If the dimensions of the front yard are 8 ft by 16 ft and the dimensions of the backyard are 24 ft by 48 ft, what is the ratio of the area of the front yard to the area of the backyard?

5. How many boxes of cookies measuring 2 inches long by 3 inches wide by 4 inches deep will fit into a box measuring 12 inches long by 15 inches wide by 16 inches deep?

6. If a tile is 8 inches by 8 inches, how many tiles are needed in a room measuring 8 ft by 12 ft?

Arithmetic Reasoning
Ratios - Post-Test with Answers

1. If there is a 9:4 ratio of cows to horses, and there are 63 cows, how many horses are there?

    cows/horses = cows/horses
    9/4 = 63/x
    Reduce 9 and 63 to become 1 and 7.
    Reducing across or up and down is possible in proportions.
    1/4 = 7/x
    x = 28 horses

2. If a woman donates $100,000 to charity and she gives her money to a school and a community center in a 3:1 ratio, how much money did she give to the school?

    Most ratio problems are solved as proportions, but this one is not. This is a different case where the amount of money is the total sum of money. Addition of the ratios is required.

    school: community center donation ratio is 3:1
    3x + x = 100,000
    4x = 100,000
    x = 25,000
    Usually when we get to x, the problem is solved.
    However, in this case, the value of x is not the final answer.
    The value of the school donation is 3x.
    3x = 3(25,000) = 75,000
    The donation to the school is $75,000.

3. If there are 20 boys and 15 girls playing in the park, what is the ratio of boys to children in the park?

    boys/total children
    boys + girls = 20 + 15 = 35
    boys/total children = 20/35 = 4/7

4. If the dimensions of the front yard are 8 ft by 16 ft and the dimensions of the backyard are 24 ft by 48 ft, what is the ratio of the area of the front yard to the area of the backyard?

Front yard   8 ft x   16 ft

                3   x   3   =   9

Backyard   24 ft x 48 ft

The ratio of the front yard to the backyard is 1:9 since the backyard is 9 times bigger than the front yard.

5. How many boxes of cookies measuring 2 inches long by 3 inches wide by 4 inches deep will fit into a box measuring 12 inches long by 15 inches wide by 16 inches deep?

2 in. x   3 in. x   4 in.
6   x   5   x   4 = 120 boxes
12 in. x 15 in. x 16 in.

6. If a tile is 8 inches by 8 inches, how many tiles are needed in a room measuring 8 ft by 12 ft?

To solve this problem, you need to know there are 12 inches in 1 foot.

Each tile is 8 in. x 8 in.
The room is 8 ft x 12 ft
Change ft. into in.
8 ft = 96 in.
12 ft = 144 in.

8 in. x   8 in.
12   x   18 = 216 tiles
96 in. x 144 in.

Arithmetic Reasoning
Tax, Tip, Commission, Overtime - Post-Test

1.   If sales tax is 8.5%, how much is the sales tax on a stroller that costs $125.00?

2.   Sally dined at a restaurant. Her bill came to $52.00, and she wants to give a 15% tip. How much should she leave for the tip?

3.   A salesperson earns $12.00 an hour and 15% commission on any sales made. If he sells $4000 in merchandise and works for 60 hours, what was his total earned for the week?

4.   Kayla earns $22.00 an hour for 40 hours and earns 1 1/2 times her salary for overtime. If she works a total of 60 hours in a week, how much will she earn that week?

Arithmetic Reasoning
Tax, Tip, Commission, Overtime - Post-Test with Answers

1. If sales tax is 8.5%, how much is the sales tax on a stroller that costs $125.00?

    To find sales tax, change percent to a decimal. Then multiply.
    8.5% = .085
    125 x .085 = 6.375 = $10.625 = $10.63

2. Sally dined at a restaurant. Her bill came to $52.00, and she wants to give a 15% tip. How much should she leave for the tip?

    To find tip, change percent to a decimal. Then multiply.
    15% = .15
    52.00 x .15 = $7.80

3. A salesperson earns $12.00 an hour and 15% commission on any sales made. If he sells $4000 in merchandise and works for 60 hours, what was his total earned for the week?

    Find the base pay by multiplying hourly rate by the hours worked.
    12 x 60 = $720.00

    Find the commission by changing the percent to a decimal and multiplying the percent by how much merchandise was sold.

    Change the percent to a decimal. Then multiply.
    15% = .15
    4000 x .15 = $600.00

    Add the base pay and the commission.
    $720.00 + $600.00 = $1320.00

4. Kayla earns $22.00 an hour for 40 hours and earns 1 1/2 times her salary for overtime. If she works a total of 60 hours in a week, how much will she earn that week?

    The quicker way to figure out overtime pay is:
    Subtract 40 hours from the total hours worked.
    60 - 40 = 20

Multiply 20 by 1.5 = 30 hours
Add 40 + 30 = 70 hours
Multiply the hourly rate by 70
22.00 x 70 = $1540.00

The more traditional way to solve overtime problems is done in the following way:

$22.00 x 40 = $880.00

Then multiply the hourly rate by 1.5 to figure out the overtime rate.
$22.00 x 1.5 = $33.00

Then subtract 40 from 60 to figure out how many overtime hours are worked.
Multiply the overtime hours by the overtime rate.
$33.00 x 20 = $660.00

Add the base pay to the overtime pay to find the total pay.

$880.00 + $660.00 = $1540.00

Arithmetic Reasoning
Interest Rate - Post-Test

1.   Lester put $900.00 in a bank account that earns 4% interest. How much interest will he earn in 3 years?

2.   Becky puts money in the bank at a rate of 2% for 4 years and earns $40.00 in interest. How much money did she put in the bank?

3.   Ryan puts $450 in the bank for 2 years and earns $45 in interest. What was the interest rate?

4.   Bill puts $2000 in the bank at a rate of 5% and earns $700 in interest. How many years did it take to earn that much interest?

5.   Jack put $800 in the bank at 5% for 2 years. The interest is compounded annually.
      Including interest, how much money will be in his bank account at the end of the 2 years?

6.   If Sarah's car is worth $10,000 and depreciates by 5% per year, how much will her car
      be worth in 2 years?

Arithmetic Reasoning
Interest Rate - Post-Test with Answers

1.   Lester put $900.00 in a bank account that earns 4% interest. How much interest will he earn in 3 years?

Interest = principal x rate x time
Change 4% to a decimal .04.
Principal = $900
Rate = 4% = .04
Time = 3 years
Interest is missing.
I = 900 x .04 x 3
Multiply 900 x 3 first and then multiply by .04 (because we want to wait until the last step to multiply the decimal to make the steps easier.)
2700 x .04 = $108.00

2.   Becky puts money in the bank at a rate of 2% for 4 years and earns $40.00 in interest. How much money does she put in the bank?

Interest = principal x rate x time
Change 2% to .02
Interest = $40.00
Rate = 2% = .02
Time = 4 years
Principal is missing.
40 = p x .02 x 4
40 = p x .08
40 = .08p
Divide by .08
40 ÷ .08 = 4000 ÷ 8 = $500

3.   Ryan puts $450 in the bank for 2 years and earns $45 in interest. What was the interest rate?

Interest = principal x rate x time
Principal = $450
Time = 2 years
Interest = $45
Rate is missing.
45 = 450 x r x 2
45 = 900r
45 ÷ 900 =
45.00 ÷ 900   = .05 = 5%

155

4. Bill puts $2000 in the bank at a rate of 5% and earns $700 in interest. How many years did it take to earn that much interest?

    Interest = principal x rate x time
    Principal = $2000
    Rate = 5%
    Interest = $700
    Time is missing.
    700 = 2000 x .05 x t
    700 = 100t
    7 = t
    7 years = time

5. Jack put $800 in the bank at 5% for 2 years. The interest is compounded annually. Including interest, how much money will be in his bank account at the end of the 2 years?

    Compound interest needs to be calculated one year at a time. The interest must be added on for each year and the formula need to be repeated for the next year and then the interest is added on again. The number of times the interest formula is used matches the number of years it was invested.

    Year 1
    I = p x r x t
    I = 800 x .05 x 1 = 40
    800.00 + 40.00 = $840.00

    Year 2
    I = p x r x t
    I = 840 x .05 x 1 = 42
    840.00 + 42.00 = $882.00

    If the answer choices on the test are not that close in value, you can find an estimate of the answer by using the simple interest formula for 2 years. The real answer will be a little bigger.
    I = 800 x .05 x 2
    I = 1600 x .05 = 80
    I = $80
    800.00 + 80.00 = $880.00
    The estimated answer would be $880.00 while the real answer is $882.00

6.   If Sarah's car is worth $10,000 and depreciates by 5% per year, how much will her car be worth in 2 years?

Year 1
$10,000 x .05 = $500
$10,000.00 - $500.00 = $9,500.00

Year 2
$9,500 x .05 = $475
$9,500.00 - $475.00 = $9,025.00

Arithmetic Reasoning
Patterns and Sequences - Post-Test

Complete the following patterns.

1.

      1, 4, 16, 64, _____

2.

      11, 20, 17, 26, 23, _____

3.

      1/3, 1/9, 1/27, _____

4.

      3, -4, 4, -3, 5, _____

5.

      1, 3, 7, 13, _____

6.

50, 41, 34, 29, _____

Arithmetic Reasoning
Patterns and Sequences - Post-Test with Answers

Complete the following patterns.

1.

   1, 4, 16, 64, _____

   Look for the pattern. In this case, each number is multiplied by 4 to lead to the next number.
   The next number would be <u>256</u>.

2.

   11, 20, 17, 26, 23, _____

   The pattern involves addition and subtraction.
   The pattern is: add 9, subtract 3, add 9, subtract 3, add 9.
   The next number would be <u>32</u>.

3.

   1/3, 1/9, 1/27, _____

   Each fraction is divided by 3 to lead to the next one.
   The denominator in each one is tripled.
   The next number would be <u>1/81</u>.

4.

   3, -4, 4, -3, 5, _____

   The pattern involves addition and subtraction.
   The pattern is: subtract 7, add 8, subtract 7, add 8, subtract 7.
   The next number would be <u>-2</u>.

5.

      1, 3, 7, 13, _____

The pattern is: add 2, add 4, add 6, add 8.
The next number would be <u>21</u>.

6.

      50, 41, 34, 29, _____

The pattern is: subtract 9, subtract 7, subtract 5, subtract 3.
The next number would be <u>26</u>.

Arithmetic Reasoning
Probability and Consecutive Integers - Post-Test

1.   If Kurt rolls a die, what is the probability of landing on a number more than 2?

2.   What is the probability of picking a Jack or a King out of a deck of cards?

3.   If there are 5 blue paper clips, 4 green paper clips, 3 yellow paper clips, and 6 red paper clips in a bag, what is the probability of picking a green or a yellow paper clip from the bag?

4.   If there are three consecutive numbers with a sum of 432, what is the largest of the three numbers?

5.   If there are three consecutive even numbers with a sum of 234, what is the second number of the three numbers?

6. If there are three consecutive odd numbers with a sum of 189, what is the smallest of the three numbers?

Arithmetic Reasoning
Probability and Consecutive Integers - Post-Test with Answers

1.   If Kurt rolls a die, what is the probability of landing on a number more than 2?

There are 6 sides on a die and 4 of them are more than 2.
The probability of rolling a die and landing on a number more than 2 is:
$4/6 = 2/3$

2.   What is the probability of picking a Jack or a King out of a deck of cards?

To solve this problem, you need to know that there are 52 cards in a deck.
There are 4 suits: hearts, diamonds, clubs, and spades.
There are 13 cards in each suit: A, 2, 3, 4, 5, 6, 7, 8, 9, 10, Jack, Queen, King.

There are 52 cards in a deck and there are 4 Jacks and 4 Kings in a deck of cards.
The probability of picking a Jack or a King out of a deck of cards is:
Jacks + Kings / Total = $4 + 4 /52 = 8/52 = 2/13$

3.   If there are 5 blue paper clips, 4 green paper clips, 3 yellow paper clips, 6 red paper clips in a bag, what is the probability of picking a green or a yellow paper clip from the bag?

There are a total of 18 paper clips, and there are 4 green and 3 yellow paper clips.
green + yellow paper clips / total paper clips = $4 + 3 / 18 = 7/18$

4.   If there are three consecutive numbers with a sum of 432, what is the largest of the three numbers?

$432 \div 3 = 144$ is the middle number
The three consecutive numbers are 143, 144, and 145.
The largest of the 3 numbers is 145.

5.	If there are three consecutive even numbers with a sum of 234, what is the second number of the three numbers?

$234 \div 3 = 78$ is the middle number.
The three numbers are consecutive even numbers so we need to skip over the odd numbers.
The three consecutive even numbers are 76, 78, and 80.
The second of the three numbers is 78.

6.	If there are three consecutive odd numbers with a sum of 189, what is the smallest of the three numbers?

$189 \div 3 = 63$ is the middle number.
The three numbers are consecutive odd numbers, so we need to skip over the even numbers.
The three consecutive odd numbers are 61, 63, 65.
The smallest of the three numbers is 61.

Arithmetic Reasoning
Distance and Rate - Post-Test

1. Two trains are in the same city and start traveling in opposite directions. If one train travels east at 40 mph and the other train travels west at 60 mph, how far apart will the trains be at the end of 4 hours?

2. If there are two planes that are 6000 miles apart and they are flying toward each other, how long will it take the two planes to meet if one is traveling at 900 mph and the other is traveling at 600 mph?

3. If two cars are traveling in the same direction, and one is traveling at a speed of 45 mph, and the other is traveling at a speed of 60 mph, how far apart will the cars be at the end of 3 hours?

4. A train travels a total distance of 640 miles. It travels at a speed of 80 mph. If the train leaves at 8 AM, when will the train reach its destination?

5. If a car travels at a rate of 50 mph for 3 hours and travels at a rate of 40 mph for 2 hours, what was the average rate of speed in miles per hour?

6. A young girl leaves home to travel cross-country and forgets her laptop. She drives at a speed of 45 mph. Her father realizes she forgot her laptop and follows her in his car 1 hour after she left. If he wants to catch up with her 4 hours after she left, how fast does he have to drive?

Arithmetic Reasoning
Distance and Rate - Post Test with Answers

1. Two trains are in the same city and start traveling in opposite directions. If one train travels east at 40 mph and the other train travels west at 60 mph, how far apart will the trains be at the end of 4 hours?

    When trains move opposite directions, add the distances each travels in an hour and multiply by the number of hours.
    $40 + 60 = 100$
    $100 \times 4 = 400$ miles

2. If there are two planes that are 6000 miles apart and they are flying toward each other, how long will it take the two planes to meet if one is traveling at 900 mph and the other is traveling at 600 mph?

    Add the distances traveled in one hour by each plane and divide into the total distance to be traveled.
    $900 + 600 = 1500$
    $6000 \div 1500 = 4$ hours

3. If two cars are traveling in the same direction, and one is traveling at a speed of 45 mph, and the other is traveling at a speed of 60 mph, how far apart will the cars be at the end of 3 hours?

    When cars travel the same direction, subtract the two speeds to see how far ahead one is after an hour. Then multiply by the total number of hours.
    $60 - 45 = 15$ mph
    $15 \times 3 = 45$ miles

4. A train travels a total distance of 640 miles. It travels at a speed of 80 mph. If the train leaves at 8 AM, when will the train reach its destination?

    $640 \div 80 = 8$ hours
    8 hours after 8 AM is 4 PM.

5.   If a car travels at a rate of 50 mph for 3 hours and travels at a rate of 40 mph for 2 hours, what was the average rate of speed in miles per hour?

$(50 \times 3) + (40 \times 2) =$
$150 + 80 = 230$ miles
The total hours traveled was $3 + 2 = 5$ hours
$230 \div 5 = 46$ mph

6.   A young girl leaves home to travel cross-country and forgets her laptop. She drives at a speed of 45 mph. Her father realizes she forgot her laptop and follows her in his car 1 hour after she left. If he wants to catch up with her 4 hours after she left, how fast does he have to drive?

The daughter leaves and travels 45 mph in the first hour. The father wants to catch up with her 4 hours after she leaves. She will have traveled 180 miles.
The father needs to cover 180 miles in 3 hours.
$180 \div 3 = 60$ mph

Steps:
1. Multiply hours by the daughter's speed.
   $45 \times 4 = 180$ miles
2. Subtract hours daughter will travel by how many hours later the father left.
   $4 - 1 = 3$ hours
3. Divide Step 1 by Step 2.
   $180 \div 3 = 60$ mph
   The father will need to travel at 60 mph.

Arithmetic Reasoning
Miles Per Gallon, Algebra in Word Problems - Post-Test

1. Felix buys gas for $2.95 a gallon. His car gets 25 miles per gallon. How much will a trip of 800 miles cost?

2. Wanda spent $108.00 on gas, and gas costs $3.60 per gallon. Her car gets 28 miles per gallon. How far did she drive?

3. Raymond and Gordon have the same number of toy cars. Raymond has 3 times as many toy cars as Kevin, and Gordon has 20 more toy cars than Kevin. How many toy cars does Kevin have?

4. Barbara is 4 years older than Dorothy, and Dorothy is 2 years younger than Nancy. If the sum of their ages is 66, how old is Dorothy?

5. A woman spends 1/5 of her income on her rent and 1/4 of the remainder of her income on food. What part of her salary does she spend on food?

6. Kimberly rode in a taxi and her trip cost $65.00. The initial mile cost $1.50, and the trip cost $.50 each additional mile. How many miles long was Kimberly's ride in the taxi?

7. Linda earns the following test grades: 95, 85, 80, and 90. What score will she need on the fifth test so that she ends up with an average of 90 in the class?

Arithmetic Reasoning
Miles Per Gallon, Algebra in Word Problems - Post-Test with Answers

1.   Felix buys gas for $2.95 a gallon. His car gets 25 miles per gallon. How much will a trip of 800 miles cost?

This problem has 2 steps in order to solve it.
To help you remember the steps, remember that in this problem there are 3 amounts mentioned: $2.95, 25, and 800.
1 of them is a money amount, and the other 2 are numbers.
Divide number into number and multiply by money.
800 ÷ 25 = 32 then multiply by $2.95
2.95 x 32 = $94.40

2.   Wanda spent $108.00 on gas, and gas costs $3.60 per gallon. Her car gets 28 miles per gallon. How far did she drive?

This problem has 2 steps in order to solve it. To help you remember the steps, remember that in this problem there are 3 amounts mentioned: $108.00, $3.60, and 28.
2 of them are money amounts, and 1 is a number.
Divide money into money and then multiply by number.
$108.00 ÷ $3.60 = 30
30 x 28 = 840 miles

3.   Raymond and Gordon have the same number of toy cars. Raymond has 3 times as many toy cars as Kevin, and Gordon has 20 more toy cars than Kevin. How many toy cars does Kevin have?

Translate into algebra.
Raymond has the same number of toy cars as Gordon.
$R = G$
Raymond has 3 times as many toy cars as Kevin.
$R = 3K$
Gordon has 20 more toy cars than Kevin.
$G = K + 20$

$R = G$
$R = 3K$ and $G = K + 20$
$3K = K + 20$

Solve for K.
2K = 20
K = 10
Kevin has 10 toy cars.

To check the work, plug back into the original equation.
R = G
R = 3K = 3(10) = 30
Raymond has 30 toy cars.
G = K + 20 = 10 + 20 = 30
Gordon has 30 toy cars.
Therefore, it does work that R = G when K = 10.

4.   Barbara is 4 years older than Dorothy, and Dorothy is 2 years younger than Nancy. If the sum of their ages is 66, how old is Dorothy?

Translate into algebra.
Barbara = Dorothy + 4
B = D + 4
Dorothy is 2 years younger than Nancy.
Dorothy = Nancy – 2
D = N – 2
This also means that Nancy is 2 years older than Dorothy.
Nancy = Dorothy + 2
N = D + 2
This formula, N = D + 2, works better for us than D = N – 2 since we need to have one variable and we want to solve Dorothy's age.
Dorothy = ?
D = ?

Barbara's age + Dorothy's age + Nancy's age = 66
(D + 4) + D + (D + 2) = 66
3D + 6 = 66
3D = 60
D = 20
Dorothy is 20 years old.

To check your work, plug 20 back in to the equation.
(D + 4) + D + (D + 2) = 66
(20 + 4) + 20 + (20 + 2) = 66
24 + 20 + 22 = 66
So Barbara is 24, Dorothy is 20, and Nancy is 22, and it checks to be correct.

5.   A woman spends 1/5 of her income on her rent and 1/4 of the remainder of her income on food. What part of her salary does she spend on food?

   After a woman spends 1/5 of her income on her rent, she has 4/5 of her income left. She spends 1/4 of her remaining money on her food.
   1/4 x 4/5 = 4/20 = 1/5

6.   Kimberly rode in a taxi and her trip cost $65.00. The initial mile cost $1.50, and the trip costs $.50 each additional mile. How many miles long was Kimberly's ride in the taxi?

   65.00 – 1.50 (for initial mile) = $63.50
   The trip cost $.50 for each additional mile.
   Division will show how many additional miles Kimberly rode in the taxi.
   $63.50 ÷ $.50 = 6350 ÷ 50 = 635 ÷ 5 = 127
   Add the first mile to the 127 miles. The total trip was 127 + 1 = 128 miles

7.   Linda earns the following test grades: 95, 85, 80, and 90. What score will she need on the fifth test so that she ends up with an average of 90 in the class?

   There are a total of 5 tests so 90 needs to be multiplied by 5 to figure out the total number of points needed to earn an average of 90.

   $95 + 85 + 80 + 90 + x = 90 (5)$
   $350 + x = 90(5)$
   $350 + x = 450$
   Subtract 350 from both sides.
   $350 + x = 450$
   $-350 \qquad -350$
   $ x = 100$

Arithmetic Reasoning
Measurement - Post-Test

1. If a 6 yd piece of wood is cut into 12 pieces, how long is each piece?

2. If 15 ft 6 inches of wood needs to be divided into 6 pieces, how long would each piece be?

3. If Betty ran 4 miles and Jackie ran 1 mile, how many more yards did Betty run than Jackie?

4. If Jacob walked 1 mile in 30 minutes, how many feet did he walk per minute?

5. If a room measures 6 ft by 9 ft, and carpet cost $4.50 per square yard, how much would it cost to carpet the room?

6. A quart of milk costs $2.00, and a gallon-sized container costs $6.00. How much would someone save by buying 3 gallons in gallon-sized containers instead of buying the same amount of milk in quart-sized containers?

7. A swimming pool has dimensions of 30 ft long by 12 ft wide by 5 ft deep. A cubic foot of water is about 7.5 gallons. How many gallons are used to fill the pool?

8. Alexis's baby was born weighing 8 lb 2 oz and Sandy's baby was born weighing 5 lb 13 oz. How much more did Alexis's baby weigh than Sandy's baby at birth?

9. Toby orders 120 pounds of hamburger meat. How many 3/4 pound hamburgers can he make using 120 pounds of meat?

10. A truck carries 2 tons of dog food and each dog receives a daily food quota of 10 ounces. How many dogs can be fed from this delivery for a total of 16 days?

11.   Justin ran 2300 meters and Craig ran 4 km. How many kilometers farther did Craig run than Justin?

12.   If a tree grows 15 mm per day, how many meters will it grow in 1 year?

13.   If Nicole ran 750 meters per day, how many km would she have run in the months of March, April, and May?

Arithmetic Reasoning
Measurement - Post-Test with Answers

1. If a 6 yd piece of wood is cut into 12 pieces, how long is each piece?

    To solve this problem, you need to know that there are 3 ft in 1 yard and there are 12 inches in 1 ft.
    Treat this as a fraction.

    6 yd/12 can be reduced to 1 yd/ 2
    6 yd/12 = 1 yd/2 = 36 inches/2 = 18 inches = 1 ft 6 inches

2. If 15 ft 6 inches of wood needs to be divided into 6 pieces, how long would each piece be?

    To solve this problem, you need to know there are 12 inches in a foot.

    15 ft 6 inches = 15 x 12 = 180 + 6 = 186 inches
    186 ÷ 6 = 31 inches or 2 ft 7 inches

3. If Betty ran 4 miles and Jackie ran 1 mile, how many more yards did Betty run than Jackie?

    To solve this problem, you need to know that there are 1760 yards in a mile.

    4 − 1 = 3 miles
    There are 1760 yards in a mile. 1760 x 3 = 5280 yards

4. If Jacob walked 1 mile in 30 minutes, how many feet did he walk per minute?

    To solve this problem, you need to know there are 5280 yards in a mile.

    There are 5280 feet in a mile.
    5280 ÷ 30 = 176 feet in a minute

5. If a room measures 6 ft by 9 ft, and carpet cost $4.50 per square yard, how much would it cost to carpet the room?

To solve this problem, you need to know that there are 3 feet in 1 yard.
The carpet cost is measured in square yards.
Change 6 ft x 9 ft into yards by dividing each one by 3.
2 yd x 3 yd = 6 square yards = 6 yd²
$4.50 x 6 = $27.00

6. A quart of milk costs $2.00, and a gallon-sized container costs $6.00. How much would someone save by buying 3 gallons in gallon-sized containers instead of buying the same amount of milk in quart-sized containers?

To solve this problem, you need to know there are 4 quarts in a gallon, and there are 12 quarts in 3 gallons.

$2.00 x 12 = $24.00
3 gallons would cost $6.00 x 3 = $18.00
$24.00 - $18.00 = $6.00 more to buy quarts instead of the gallon-sized containers.
A person would save $6.00 by buying the gallon-sized containers.

7. A swimming pool has dimensions of 30 ft long by 12 ft wide by 5 ft deep. A cubic foot of water is about 7.5 gallons. How many gallons are used to fill the pool?

To solve this problem, you need to know the volume of a rectangular box.
V = L x W x H

Find the volume of the pool.

L x W x H
30 x 12 x 5 = 1800 cubic ft = 1800 ft³
Multiply the volume by the number of gallons per cubic foot.
1800 x 7.5 = 13,500 gallons

8. Alexis's baby was born weighing 8 lb 2 oz, and Sandy's baby was born weighing 5 lb 13 oz. How much more did Alexis's baby weigh than Sandy's baby at birth?

To solve this problem, you need to know there are 16 ounces in 1 pound.

179

Julie A. Hyers ©2020

There are two ways to solve this problem.
a. Change the pounds to ounces and subtract.

8 lb 2 oz = 8 x 16 = 128 + 2 = 130 oz
5 lb 13 oz = 5 x 16 = 80 + 13 = 93 oz
130 − 93 = 37 oz
37 ÷ 16 = 2 lb with 5 oz left over
2 lb 5 oz

b. Borrow from the pounds to be able to subtract the ounces.

8 lb 2 oz − 5 lb 13 oz
Since you cannot take 13 oz away from 2 oz,
you need to turn 8 lb into 7 lb and add 16 oz to the ounces.
7 lb 18 oz − 5 lb 13 oz = 2 lb 5 oz

9.   Toby orders 120 pounds of hamburger meat. How many 3/4 pound
       hamburgers can he make using 120 pounds of meat?

       Toby has 120 pounds of hamburger meat that needs to be divided into 3/4
       pound hamburgers.

       120 ÷ 3/4 =

       120/1 x 4/3 = 480/3 = 160 hamburgers

10.   A truck carries 2 tons of dog food and each dog receives a daily food quota of
       10 ounces. How many dogs can be fed from this delivery for a total of 16 days?

       To solve this problem, you need to know that:
       1 ton = 2000 pounds and 1 pound = 16 ounces
       2 tons = 4000 pounds

       Each dog eats 10 ounces a day, which is 160 ounces in 16 days.
       160 ounces = 10 pounds
       4000 ÷ 10 = 400 dogs
       or
       4000 pounds = 64,000 ounces
       10 ounces a day = 160 ounces in 16 days
       64,000 ÷ 160 = 400 dogs

11.   Justin ran 2300 meters, and Craig ran 4 km. How many kilometers farther did Craig run than Justin?

To solve this problem, you need to know there are 1000 meters in a kilometer.

Justin ran 2300 meters, and Craig ran 4 km, which equals 4000 meters.
4000 − 2300 = 1700 meters = 1.7 km
or
Justin ran 2.3 km, and Craig ran 4 km.
4.0 km − 2.3 km = 1.7 km

12.   If a tree grows 15 mm per day, how many meters will it grow in 1 year?

To solve this problem, you need to know that there are 1000 mm in a meter.
You also need to know that there are 365 days in a year.

15 mm x 365 = 5,475 mm
There are 1000 mm in a meter.
5,475 ÷ 1000 = 5.475 meters

13.   If Nicole ran 750 meters per day, how many km would she have run in the months of March, April, and May?

To solve this problem, you need to know there are 1000 meters in a kilometer.

You also need to know how many days are in each month of the year.

January = 31 days
February = 28 days, except on Leap Year when it has 29 days once every 4 years.
March = 31 days
April = 30 days
May = 31 days
June = 30 days
July = 31 days
August = 31 days
September = 30 days
October = 31 days
November = 30 days
December = 31 days

There is also an old poem to help remember the days in the month.

Thirty days hath September, April, June, and November.
All the rest have thirty-one.
February has twenty-eight,
But leap year coming one in four.
February then has one day more.

March has 31 days, April has 30, and May has 31 days.
31 + 30 + 31 = 92 days
92 x 750 = 69,000 meters
There are 1000 meters in a kilometer
69,000 ÷ 1000 = 69 km

Made in the USA
Columbia, SC
25 August 2024